Report of the

EXPERT CONSULTATION ON IDENTIFYING, ASSESSING AND REPORTING ON SUBSIDIES IN THE FISHING INDUSTRY

Rome, 3–6 December 2002

FOOD AND AGRICULTURE ORGANIZATION OF THE UNITED NATIONS
Rome, 2003

ISBN 92-5-104890-8 √

© FAO 2003

PREPARATION OF THIS DOCUMENT

This document is the final report of the Expert Consultation on Identifying, Assessing and Reporting on Subsidies in the Fishing Industry which was held in Rome from 3 to 6 December 2002.

Distribution:

All FAO Members
List of participants
Other interested Nations and national and international organizations
FAO Fisheries Department
FAO Regional Fishery Officers

iv

FAO.
Report of the Expert Consultation on Identifying, Assessing and Reporting on Subsidies in the Fishing Industry. Rome, 3-6 December 2002.
FAO Fisheries Report. No. 698. Rome, FAO. 2003. 81p.

ABSTRACT

The Expert Consultation on Identifying, Assessing and Reporting on Subsidies in the Fishing Industry met at FAO Headquarters for four days starting 3 December 2002. Fourteen experts, from as many countries, invited in their individual capacity, attended the Consultation. They elected Dr Trond Bjorndal as the Chair and Ms Sita Kuruvilla as vice-chair and adopted an agenda that included two main issues: (i) the review of a draft guide for identifying, assessing and reporting on fishery subsidies; and, (ii) the comparison of the magnitude and effects of subsidies in the fisheries sector.

The experts found that, with some few modifications, the draft guide constituted an excellent support for subsidy studies and recommended that it be used. To further expand the usefulness of the guide the experts recommended that FAO work on: (i) the long-term effects of subsidies, (ii) resource pricing; and, (iii) the effects of government inaction. They also adopted eight conventions for using the guide. These conventions are intended to facilitate the international comparisons of subsidy studies.

The experts considered it important that available methodologies be used to assess impacts flowing from actions taken by recipient of subsidies on environment, trade, economic growth and social conditions. The Consultation concluded that FAO should promote the development and use of appropriate models for the evaluation of impacts of subsidies through actual case studies.

TABLE OF CONTENTS

INTRODUCTION

1. The Expert Consultation on Identifying, Assessing and Reporting on Subsidies in the Fishing Industry met in FAO Headquarters, Rome, Italy from 3-6 December 2002.

2. The Expert Consultation was attended by 14 experts and 3 observers. They are listed in Appendix C. The documents placed before the experts are listed in Appendix B.

OPENING OF THE CONSULTATION

3. The Expert Consultation was opened by Mr Ichiro Nomura, the Assistant Director General of Fisheries. In welcoming the participants Mr Nomura noted that subsidies to the fishing industry continue to be a concern in international meetings on fisheries and on the environment. Therefore it is understandable that FAO Members want to find out more about this controversial policy instrument. He went on to describe some of the work carried out during the last two years, providing some details about the background to the development of the draft Guide on identifying, assessing and reporting on subsidies in the fishing industry. He encouraged the experts to work hard to improve the Guide and to advise the Organization on how to expand the work to include also the evaluation of impacts on sustainability, trade and development. Mr Nomura's Opening Statement is attached as Appendix D.

ELECTION OF CHAIRPERSON

4. The Expert Consultation elected Dr Trond Bjorndal as its Chair and Ms Sita Kuruvilla as its vice-Chair.

ADOPTION OF THE AGENDA AND TIMETABLE

5. The Expert Consultation adopted the agenda and timetable as contained in Appendix A to this report.

BACKGROUND AND PREPARATIONS

6. In trying on the one hand to implement the recommendation made by the FAO Committee of Fisheries (COFI) in 2001 and, on the other hand, to address the conclusions of the first FAO Expert Consultation on fishery subsidies[1] it became apparent that the lack of detailed data on the nature, magnitude and impact of subsidies on the sector constituted a formidable obstacle to progress. The FAO Fisheries Department then decided as a first technical step, to try to develop a methodology for generating quantitative data of relevance for assessing the impact of subsidies. A Guide helping to identify, assess and report on subsidies – without going into their potential secondary effects outside the recipient enterprises – could constitute the first step in a strategy that in a second and subsequent step would help to develop quantitative indicators of the importance of these secondary effects (on trade, environment and development) as well as facilitate their analysis.

---

[1] Expert Consultation on Economic Incentives and Responsible Fisheries (Rome, 28 November-1 December 2000) - FAO Fisheries Report No.638

7. In the course of 2002, FAO Fisheries Department of FAO developed a draft "Guide for Identifying, Assessing and Reporting on Subsidies in the Fisheries Sector" (the Guide). The work was undertaken as follows:

- a first preliminary draft was prepared;
- approval was obtained to test the draft Guide by conducting a study of fishery subsidies in four Member Countries (a transition economy country, a low income country, an upper middle income country and a developed country);
- letters of agreements were signed with offices/organizations in the concerned countries for the execution prototype studies;
- FI provided through missions to the field technical backstopping and supervision to the contracted national organizations/institutions in the use of the draft Guide for the preparation of prototype studies[2];
- the draft Guide was rewritten based on the experience derived from the prototype studies; and
- the second version of the draft Guide was peer-reviewed and the peer review became the basis for the third version submitted to this Consultation.

REVIEW OF "GUIDE FOR IDENTIFYING, ASSESSING AND REPORTING ON SUBSIDIES IN THE FISHERIES SECTOR"

8. The Consultation decided to conduct the discussion according to the suggestions provided in document: "Guide for discussing of the draft Guide for identifying, assessing and reporting on subsidies in the fisheries sector" (FI:ECFS/2002/3). The Guide is reproduced in Appendix E.

9. Ms Lena Westlund, principal author of the Guide, at the invitation of the Chair, made a presentation of issues that emerged in the course of developing the Guide. This was followed by a general discussion of the context and purpose of the Guide. The following paragraphs summarize the observations made and conclusions reached in the course of these discussions.

10. Subsidies are not studied in a vacuum. It is important that the data and information that is collected about subsidies be selected so that it becomes useful not only for fulfilling the analysis specified in the guide but also so that it is amenable for use in studies aiming to assess and evaluate the impact of subsidies on sustainability, economic development and trade. This has implications also for the way information is presented

11. However, at the same time the Consultation recognized the fact that the Guide is not normative in respect of subsidies. While it provides guidance for assessing the costs to the provider and the benefits to the receiver, it provides no methodologies for evaluating whether subsidies have impacts on social, economic, environmental, trade or other characteristics of the economies in which they are used. Information generated by using the Guide would need to be further evaluated – and complemented by other types of information – before judgements about impact can be made.

---

[2] The report resulting from the study carried out in Trinidad and Tobago is attached as Information document FI:ECFS/2002/Inf. 4 - Study on Subsidies in The Fisheries Sector of Trinidad and Tobago

12.   The Guide provides a framework for how to define subsidies but leaves the user free to select which government policies to consider as subsidies and which not to consider as subsidies. There was a wide consensus amongst the experts that this is not a constructive approach. Two remedies were suggested. First, it was suggested that the Guide should also make clear to its users what, from the point of view of economic theory, could or ought to be considered as a subsidy and to facilitate international comparisons. Second, the Guide should also provide information as to what should be considered as a subsidy to the fisheries sector, given international practice. It was recognized that international practice is influenced by institutional arrangements as well as by public sector resource availability and that any international comparisons must recognize such differences.

13.   The Consultation also considered the question of international comparisons of the magnitude of subsidies and their assessment. It recognized that both in respect of criteria for identifying subsidies, as well as for benchmarks, to assess their values, more work needs to be done. Possibly the task will be easier in respect of agreeing to criteria for identifying subsidies than to agree on benchmarks for their assessments.

14.   The Consultation reviewed the concepts of 'positive versus negative' and 'good versus bad' subsidies. It noted that amongst economists there is a shared understanding of the connotation of these terms. But, as that is not the case for those who deal with this issue and are not of the economic profession, it was agreed to try to avoid using these terms and, where useful, replace the "positive versus negative" terminology with 'revenue enhancing/decreasing and 'cost enhancing/decreasing'.

15.   The Consultation also noted that it may not be sufficient to note the effect on the recipient only. In order to get a grasp of the total outcomes of a policy it is necessary to look also at the economic effects on the industry and on society as a whole.

16.   In discussing Chapter 4 of the Guide (What is a fisheries subsidy?) the Consultation noted that the very wide set of public policies considered, meant that the *de facto* definition of subsidies was much wider in scope than that applied in the WTO Agreement on Subsidies and Countervailing Measures. This was noted with approval as the Consultation held the view that this wider definition is useful for subsequent evaluation of impacts on the environment.

17.   In Chapter 4, the Guide makes use of the concept of "specificity" in identifying if a policy measure is a subsidy or not. The Guide designates a policy measure as 'specific" if it is not "normal". Some experts felt that common sense, economic theory and analysis show that some policy measures – although applicable to a whole economy – in fact have impacts on recipients that eventually lead to effects that are undesirable. Therefore such policy measures, although normal in the economy concerned, could be considered as subsidies.

18.   It was pointed out, and acknowledged by the Consultation, that in many ways developing countries represent a special situation and that this must be considered when evaluating the impacts of subsidies.

19.   Furthermore the Consultation agreed that support to the fishery sector funded through foreign aid should be considered as subsidy.

20.   The Consultation had an extensive discussion of the Guide's four categories of subsidies (Chapter 5): (1) direct financial transfers; (2) services and indirect financial

transfers; (3) interventions with different short and long-term effects; and, (4) lack of intervention. While it was recognized that the first two categories are easier to work with, it was agreed as important to keep the four categories in the Guide. Furthermore it was concluded that:

- subsidies included in categories 1 and 2 also can have long-term effects
- subsidies in categories 1 and 2 generally can be measured in monetary terms.

21.   There was also general agreement that the Guide should not be rigid in the attempt to classify subsidies into different categories. The categories are meant to be helpful to those using the Guide. Therefore the purpose of the categories should be presented clearly in the Guide.

22.   It was agreed that it is useful to split category 2 into sub-categories of which there should be at least two; direct transfers and services (including, *inter alia,* costs of fishery management). On the other hand, no definite advice could be given about the maximum duration of the concept "short-term". It was felt more appropriate to let the circumstances decide this limit. The Consultation could not identify a general rule for separating direct effects from second-stage effects and indirect consequences.

23.   The Consultation agreed that impact on the economic results of up-stream or down-stream activities – should not be considered by the Guide. The indirect consequences – in the shape of externalities – should be identified but not assessed or quantified through the Guide. This should be addressed in the follow-up work attempting to assess the impacts of subsidies on environment, development and trade.

24.   In the course of the discussion of Chapter 6 (Assessing subsidies) it became clear that with respect of the period covered in the identification of subsidies most of the experts in the Consultation favoured the approach taken in the Guide, that is to calculate the yearly volume, or value, of subsidies. In addition, the long-term effects must be addressed to the degree this is possible.

25.   The Consultation also noted that in respect of infrastructure (in particular harbours and similar large fixed assets) and other services, the procedure for estimating the value to the user, recipient must be very carefully done. This means, *inter alia,* that it is not the intended purpose of a facility that is important – it is its actual use. For example a harbour built for fishing vessels – that turns out to be used exclusively by pleasure craft – will not have had any effects on the fishing industry and should therefore not be considered as a subsidy.

26.   In respect of infrastructure, it was agreed that in the Guide it would be useful to reserve the term "infrastructure" for physical infrastructure only and should not be used to describe institutions and government services.

27.   With respect to opportunity costs of the government resources used, it was agreed that in respect of major government supplied infrastructure (harbours) and services (loans) the opportunity cost of capital should preferably be calculated.

28.   Free or below market price access to fishing grounds was discussed extensively. The issue is complicated by: (i) the distinction between cost-recovery of fishery management costs

and a payment for the access to the resource, and by (ii) the issue of what is a normal access fee.

29.   The Consultation agreed that the starting point for the discussion is that the "normal" case is that management costs and research and development costs should be recovered and, if they are not, then this amounts to a subsidy provided to the sector. In this context, however, the Consultation noted that when the value of this subsidy to the industry is assessed care should be taken.

30.   The Consultation then discussed what could be a norm for estimating what can be expected to be a reasonable "pure" fee for the access, which is a fee that does not include payment for management services. It noted that such a fee will be a portion of the resource rent. However, for practical reasons the Consultation quantified the access fee as a percentage of the landed value of catch, not as a percentage of the resource rent. The Consultation therefore suggested that the Guide should:

- where possible investigate each fishery to see what resource rents are generated;
- where no information could be obtained, a range of 3–5 % of the landed value could be used in the first instance as an indicator of the appropriate use fee and of the value of the subsidy provided if not collected
- finally, the Guide should acknowledge that for some fisheries a zero access fee would be appropriate, and therefore no resource rent subsidy is involved.

31.   The Guide gives some examples of the difficulties encountered when government facilities or services are used also by other firms than those in the fishery sector. In the Guide, allocation keys are the ratios used to divide the costs to the producer – and the benefits to the recipients – between the fisheries sector and the rest of the economy. Allocation keys were discussed. The Consultation recognized that the users of the Guide would have to be pragmatic in respect of defining allocation keys.

32.   The Consultation, after some discussion concluded that it is reasonable to value access to resources in foreign countries differently from access to resources in national waters.

33.   The Consultation recognized that assessing category 3 and category 4 subsidies is a difficult task. It may be particularly complicated in fisheries exploited by more than one country and in those subject to international fisheries agreements. The advice provided in the Guide in respect of these categories of subsidies should be strengthened. It would be useful to undertake analytical work aiming to clarify how subsidies in categories 3 and 4 impact on recipients.

34.   The Consultation reviewed Chapter 7 of the Guide; "costs and earnings analysis – the impact of industry profits". It found that costs and earnings information concerning subsidy recipients is needed. Normally such information will be presented on an industry basis. It may be useful to consider how the annual information developed through the use of the Guide, can be placed into the context of a longer period.

35.   In reviewing Chapter 8 of the Guide - comparative analysis – the Consultation noted that the ratios proposed in the Guide would have a national and an international use. The ratios provided by the guide are intended for use in a national context. It was agreed that further work would be needed before these rations could be used in international comparisons.

36.   Before the Consultation concluded its discussion of the Guide, Ms Sita Kuruvilla described how an earlier version of the Guide had been used by the Government of Trinidad and Tobago to identify, assess and report on subsidies. The Consultation noted the substantial amount of work done and the comprehensive information produced and commended the government for the effort.

37.   In concluding its review of the draft "*Guide for identifying, assessing and reporting on subsidies in the fisheries sector*", the Consultation expressed satisfaction at the excellent work done in developing the Guide. It noted that it will be useful for the various Intergovernmental Organizations that do work in this area to agree on a common method and reporting format. It was further recognized that the Guide is limited in scope as it does not permit the user to assess the degree to which subsidies reach the objectives established by governments, as those objectives go beyond effects on the economic results of recipient enterprises.

COMPARING THE MAGNITUDE AND THE EFFECTS OF SUBSIDIES IN THE FISHERIES SECTOR

38.   The Consultation considered how to promote a standardized use of the Guide. It reviewed a number of conventions aiming to facilitate the comparison of studies based on the Guide. The conventions concern the use of methodology in the study as well as the report produced as a result of the study.

39.   The Consultation agreed to propose the following conventions:

> *First convention*: specify geographical and sub-sector scope of the study;
> *Second convention*: list subsidies that have been identified and analyzed in the study;
> *Third convention*: specify the benchmarks used for quantifying subsidies;
> *Fourth convention*: allocation keys for joint subsidies to be clearly specified in the study;
> *Fifth convention*: include administrative costs incurred by the provider as part of the cost of the subsidy to the provider;
> *Sixth convention*: specify when opportunity costs to the subsidy provider have been included in the estimate of subsidies;
> *Seventh convention*: base the value of direct financial transfers on the actual government expenditure – depreciated over time when appropriate – and the financial costs that the recipients may have avoided by the receiving the transfer.
> *Eighth convention*: consider goods and services provided to the recipient to have a subsidy value corresponding to the difference between what the recipient would have paid for the equivalent goods and/or services if provided in the market and what he/she in fact paid to the public provider.

40.   The Consultation then turned to the question of how to evaluate the various impacts of subsidies.

41.   The document: "What makes a subsidy environmentally harmful: developing a checklist based on the conditionality of subsidies" was presented to the Consultation by Mr Anthony Cox, of the OECD secretariat. It was pointed out that the checklist is intended to be used in several economic sectors including fisheries, that it is a tool which may be used to rank subsidies according to their environmental effects but that it does not substitute in-depth

study which would be needed to thoroughly document those effects. In fact, already in the OECD Workshop where the proposal was first made, some modifications had been made to the checklist which would make it more appropriate for fisheries. The Consultation was informed that in respect of fisheries, the OECD secretariat will now go ahead with case studies and with work on making the "policy filter" concept more explicit and precise.

42. The Consultation recognized the potential usefulness of the checklist but considered it important to work directly on applying available methodologies to assessing the impacts created by the actions of recipients of subsidies on the environment, trade, economic growth, social conditions.

43. The Consultation furthermore considered that, if FAO decides to develop a checklist for ranking subsidies according to their impacts, this work should be undertaken in close Consultation with relevant international organizations.

CONCLUSIONS AND RECOMMENDATIONS

44. In order to avoid duplication of work and provide the occasion for synergies to occur amongst Intergovernmental Organizations addressing issues related to the use of subsidies in fisheries, the observers were invited to inform the Consultation of their relevant programmes of work.

45. The observer from the CPPS, Mr Adolfo Jalil, stressed the importance that the CPPS attaches to work aiming to provide information of the impacts of subsidies on trade and environment. The CPPS is promoting work on these matters through recently established working groups involving member countries. He stated that on this subject the CPPS would like to establish a framework for technical collaboration with FAO.

46. The observer from the OECD, Mr Anthony Cox, referred to the long history of OECD's work on subsidies and fisheries and mentioned that the last in this series of work are a study on trade liberalization and fisheries, and a study on fisheries management costs. As a follow-up to the recent OECD Workshop on Environmentally Harmful Subsidies, the fisheries secretariat will undertake case studies using the checklist (see paragraph 41) and work on the associated policy filter.

47. The observer from UNEP, Ms Anja von Moltke, stated that UNEP's work on fisheries has gained importance as a result of the Doha Ministerial Declaration. Part of the present work programme tries to document the interaction between subsidies, overcapacity and overfishing. Several country studies and informal workshops involving stakeholders including relevant organizations were conducted to address these issues. Future work will focus on analyzing the impacts of subsidies under different management conditions and providing guidance to countries for the development of sustainable fisheries policies.

48. Following these presentations the Consultation addressed the question of what would be appropriate follow-up and continuation on subsidies and fisheries.

49. The Consultation concluded that the Guide is a most appropriate tool to use in the study of subsidies and is ready for use. The experts encouraged FAO to assist Members in undertaking studies based on the Guide and to make those reports public. Finally, it was felt

to be important that FAO be informed of the experiences derived in its use as, at some time later, it may be appropriate to undertake a revision of the Guide.

50.   Simultaneously, FAO should work to improve guidance on how to identify and assess subsidies classified in categories 3 and 4. Initially this work might aim to provide quantitative assessments. Also, it is important that guidance be provided on how to place the annual analysis into a multi-year framework.

51.   The Consultation specifically urged FAO to work on the following issues: (i) the long-term effects of subsidies; (ii) resource pricing, and (iii) the effects of government inaction.

52.   The Consultation recognized that the data generated by the application of the Guide can be used to carry out empirical analyses for the purpose of estimating impacts. Essentially it is a matter of determining the stimulus to the firm, assessing the firm's reaction to that stimulus, and then determining the short-term and long-term effects on fishing capacity, on fish stocks, trade, etc. The Consultation strongly recommended that this type of analysis be undertaken.

53.   The Consultation concluded that FAO should promote development and use of appropriate models for the evaluation of impacts of subsidies through actual case studies.

54.   Regarding environmental impacts, various methods are available. They include applying econometric and bio-economic models. Models will need to be adjusted to handle different kinds of subsidies, *inter alia* management costs as well as the dynamics of fishing effort and fish stocks. The Consultation recognized that considerable effort will be needed in terms of model development and data collection. It is unlikely that one model can be used to assess all possible types of impacts, rather there will be a need to tailor models to the type of impact studied.

55.   Regarding the impacts on trade, a different model approach will be required. Global competitiveness analysis, as used in the international trade literature, allows immediate use of subsidy data (in conjunction with data on input-output coefficients for harvesting and processing activities) to assess the impact of subsidies on fisheries trade. Because of its partial equilibrium approach, it is relatively easy to perform comparative static or sensitivity analysis around important parameters of fisheries activities.

56.   The Consultation stated that case studies, based on models outlined above, would be appropriate to analyse both the environmental impact and the impact on trade of subsidies and recommends such case studies be undertaken.

ADOPTION OF THE REPORT

57.   The report was adopted on 6 December 2002.

AGENDA

(a)   Opening of the Expert Consultation

(b)   Administrative arrangements for the meeting

(c)   Election of Chairman

(d)   Review of "Guide for identifying, assessing and reporting on subsidies in the fisheries sector"

(e)   Comparing the magnitude and the effects of subsidies in the fisheries sector.

(f)   Conclusions and recommendations

(g)   Adoption of the report

LIST OF DOCUMENTS

FI:ECFS/2002/1	Provisional Agenda
FI:ECFS/2002/2	Introduction to the Draft Guide for Identifying, Assessing and Reporting on Subsidies in the Fisheries Sector
FI:ECFS/2002/3	Assessing and Reporting on Subsidies in the Fisheries sector Guide for discussion of the Draft Guide for Identifying
FI:ECFS/2002/4	Draft guide for Identifying, Assessing and Reporting on Subsidies in the Fisheries Sector
FI:ECFS/2002/5	Comparing the magnitude and the effects of subsidies in the fisheries sector
FI:ECFS/2002/Inf.1	List of documents
FI:ECFS/2002/Inf. 2	List of participants
FI:ECFS/2002/Inf. 3	Prospectus
FI:ECFS/2002/Inf. 4	Study on Subsidies in the Fisheries Sector of Trinidad & Tobago
FI:ECFS/2002/Inf. 5	Report of the Expert Consultation on Economic Incentives and Responsible Fisheries (Rome, 28 November-1 December 2000) FAO Fisheries Report No.638
FI:ECFS/2002/Inf. 6	Papers presented at the Expert Consultation on Economic Incentives and Responsible Fisheries (Rome, 28 November-1 December 2000) FAO Fisheries Report No.638/Suppl.
FI:ECFS/2002/Inf. 7	What makes a subsidy Environmentally Harmful: Developing a checklist based on the conditionality of subsidies. Author: Jan Pieters. Presented at the OECD Workshop on Environmentally Harmful Subsidies (Paris, 7-8 November 2002)
FI:ECFS/2002/Inf. 8	Report of the Ad Hoc Meeting of Intergovernmental Organizations on Work Programmes Related to Subsidies in Fisheries (Rome, 21-22 may 2001). FAO Fisheries Report No. 649
FI:ECFS/2002/Inf. 9	Report of the Second Ad Hoc Meeting of Intergovernmental Organizations on Work Programmes Related to Subsidies in Fisheries (Rome, 4-5 July 2002). FAO Fisheries Report No. 688

LIST OF PARTICIPANTS

CAMEROON

NJIFONJOU, Oumarou
Fisheries Economist
SRHOL-IRAD, PMB 77 Limbe
Cameroon
Tel.: (237) 9987616
Fax: (237) 3332376
E-mail: njifonjo@caramail.com

CANADA

SCHRANK, William E.
Professor
Department of Economics
Memorial University
St. John's, Newfoundland, A1C 5S7
Canada
Tel. 902,245 6749
Fax: 902.245 5141
E-mail:wschrank@mun.ca;
w.schrank@ns.sympatico.ca

CHINA

ZHANG, Linxiu
Professor and Deputy Director
Center for Chinese Agricultural Policy
(CCAP)
Chinese Academy of Sciences (CAS)
Building 917, Datun Road, North Asian
Games Village,
Beijing 100101, China
Tel. (86.10) 64856834/64889440
Fax: (86.10) 64856533
E-mail: lxzhang@public.bta.net.cn

ICELAND

MATTHIASSON, Thorolfur Geir
Associate Professor
Faculty of Economics and Business
University of Iceland, Iceland
Tel. (354)5671510/(354)525 4530
Fax: (354)5526806
E-mail: totimatt@hi.is

INDIA

DATTA, Samar K.
Professor and Chairman
Center for Management in Agriculture
(CMA)
Indian Institute of Management (IIM)
Vastrapur, Ahmedabad 380015
Gujarat, India
Tel. (91)796324818
Fax: (91)796306896
E-mail: sdutta@iimadh.ernet.in;
samardatta@hotmail.com

INDONESIA

JUSUF, Gellwynn
Advisor, Ministry of Marine Affairs and
Fisheries
Gd. Humpuss. Medan Merdeka
Timur No. 16, Jakarta 10110
Indonesia
Tel. (62-21) 3522515
Fax : (62-21)3522515
E-mail: gellwynn@cbn.net.id

JAPAN

KASE, Kazutoshi
Up to March 2003:
Visiting Fellow
Nissan Institute of Japanese Studies
27 Winchester Road, Oxford, OX26NA,
United Kingdom
Tel. +441865284515
Fax +441865274574
E-mail: kazutoshi.kase@st-antonys.oxford.ac.uk
After March 2003:
Professor
Institute of Social Sciences
University of Tokyo
3-7-1 Hongo, Bunkyo-ku, Tokyo
113-0011 Japan
Tel: (81.3)58414976
Fax: (81.3)5841 4905
E-mail: kase@iss.u-tokyo.ac.jp

NEW ZEALAND

SHARP, Basil Milsom H.
Professor
Dept. of Economics
University of Auckland
PB 92019 Auckland
New Zealand
Tel. 64-9 373 75 99
Fax: 64-9 373 74 27
E-mail: b.sharp@auckland.ac.nz

NORWAY

BJORNDAL, Trond
Professor
Centre for Fisheries Economics
Norwegian School of Economics and
business Administration
Helleveien 30, N-5045 Bergen
Norway
e-mail: Trond.Bjorndal@nhh.no
t.bjorndal@ic.ac.uk

PERU

ZUZUNAGA, Jorge
Fisheries Advisor
Vice-Ministerio de Pesquería
Ministerio de la Producción
Calle Uno Oeste No. 060 – Urbanización
Corpac, San Isidro, Lima
Peru
Tel.: (511)4753218/(511)2243416
Fax: (511)2242950
E-mail: jzuzunaga@minpes.gob.re
jzuzunag@yahoo.com

SIERRA LEONE

KAINDANEH, Peter Munda
Economist
1301 Walkers Line
Burlington ON, L7M 4N7
Canada
Tel.: 905 332 3305
Cell. Tel.: 905 315 0269
Fax: 905 332 7934
E-mail: kaindaneh@yahoo.com

TRINIDAD AND TOBAGO

KURUVILLA, Sita Heidi
Fisheries Officer
Fisheries Division
Ministry of Agriculture, Land and Marine
Resources, St Clair Circle
Port of Spain
Trinidad and Tobago
Tel. +18686271684 (Personal)/
(186)86344504/5 (Office)
Fax: +18686344488
E-mail: tsk@tstt.net.tt (Personal)
Mfau2fd@tstt.net.tt (Office)

SPAIN

FRANQUESA, Ramón
Professor
Gabinete de Economía del Mar
Facultad de Ciencias Económicas
Universidad de Barcelona
Av. Diagonal 690
08031 Barcelona
España
Tel. +34932856 803
E-mail: ramon@gemub.com

UNITED STATES OF AMERICA

MILAZZO, Matteo
Policy Analyst
National Marine Fisheries Service
NOAA, US Department of Commerce
1315 East-West Highway, Silver Spring,
MD 20910.
USA
Tel. (301) 713-2276
Fax: (301) 713-2313
E-mail: Matteo.Milazzo@noaa.gov

FAO

WIJKSTRÖM, Ulf N.
Chief, Development Planning Service
Fisheries Policy and Planning Division
Tel: (+39.06)57053156
Fax: (+39.06)57056500
E-mail: Ulf.Wijkstrom@fao.org

GUMY, Angel
Senior Fisheries Planning Officer
Development Planning Service
Tel: (+39.06)57056471
Fax: (+39.06)57056500
E-mail: Angel.Gumy@fao.org

WESTLUND, Lena

Consultant
Badhusvägen 13
132 37 Salts Jö-Boo
Sweden
Tel: (+46)8 57028750
Mobile: (+46)708 548813
E-mail: lena.westlund@swipnet.se

OBSERVERS

Anja **VON MOLTKE**
Economics and Trade Branch
Division of Technology, Industry and
Economics
United Nations Environment Programme
International Environment House
15 Chemin des Anémones -CH-1219
Genève, Switzerland
Tel: 41-22-917 8137
Fax: 41-22-917 8076
E-mail: anja.moltke@unep.ch

Anthony **COX**
Senior Analyst
Fisheries Division
OECD
2 rue Andre-Pascal
75775 Paris Cedex 16, France
Tel: (33) 1 45 24 95 64
Fax: (33) 1 44 30 61 21
email: anthony.cox@oecd.org

Alfonso Alvarez **JALIL**
Economic Director - Comisión Permanente
del Pacífico Sur (CPPS)
Edificio Inmaral, 1er piso
Av. Carlos Julio Arosemena, Km. 3
Guayaquil, Ecuador
Tel.: (593) 4 2221 202/2221 203
Fax: (593) 4 2221 201
E-mail: direcono@cppsnet.org
Web: www:cpps_int.org

OPENING STATEMENT BY
MR ICHIRO NOMURA, ASSISTANT DIRECTOR-GENERAL
FAO FISHERIES DEPARTMENT

Welcome to Rome; it is a pleasure to see you here. And, thank you for accepting our invitation to join the FAO Expert Consultation on Identifying, Assessing and Reporting on Subsidies in the Fishing Industry.

In recent years, as you are well aware, subsidies have been much discussed in international meetings dealing with fisheries and/or the environment. So it is not surprising that several FAO Members would like to know more about subsidies, particularly as they constitute a controversial group of policy measures. So Governments have a natural interest in finding out just how effective a policy measure they are.

Two years ago, the FAO Committee of Fisheries advised us here at Headquarters to continue to study subsidies in the fishery sector; the Committee asked us to undertake qualitative and quantitative assessment of them and of their impacts.

This sounds deceptively simple. But, as you know, it is not. The number of fisheries to be studied is very large, and – worse – we do not have a proven tool to use. So we set out by trying to develop the tool, a tool that should be sharp yet universal (a type of Swiss knife) and not too demanding to use. The availability of such a tool should make it possible for any government to get a rough idea of the costs and benefits of economic policy measures - including subsidies - applied to the fishery sector.

A rough version of this tool has been developed. The artisan is Lena Westlund. You all will come to know her as Ms Westlund will be here throughout the Expert Consultation. Ms Westlund– and my colleagues in the Fisheries Department – has dedicated much effort to come up with a guide that is both practical and useful. In the course of this year Ms Westlund has spear-headed four prototype studies of the guide; one each in four countries representing somewhat different economies.

One of the countries was Trinidad and Tobago. And, I take this opportunity to thank the Government of Trinidad and Tobago – through Ms Kuruvilla who is with us - for having permitted us to reproduce the report of the prototype study for the benefit of this Consultation.

However, the guide is not all that is needed to quantify subsidies in the fisheries sector, but it is a first step. It needs to be used and then improved based on the accumulated experience of users. But before submitting it to a wider audience to use, we want you to apply your expertise to it and thereby help us improve it.

The guide is not all that is on the agenda. You will also be asked to consider how to develop practical methods for measuring – at least in a qualitative manner – the effects of subsidies on environment, trade and development.

This part of the Expert Consultation takes as its starting point a check list developed by a Dutch economist, Mr Jan Pieters. Mr Pieters presented the check list last month in an OECD workshop in which participants discussed environmentally harmfully subsidies. The OECD secretariat has most kindly agreed to let us use the document – and the associated power point presentation – in this Expert Consultation. For this we are most grateful.

Thus in the second part of the Expert Consultation we want your advise about directions for future work. I expect you will get to this task sometime late tomorrow or early on Thursday. I do not expect you at that time to attempt to finalize that check list for application in fisheries. Of course it would be marvelous if you could, but perhaps it is more realistic for you to discuss its strength and weaknesses and then to formulate your considered view of whether or not we should proceed to work with the check-list as a starting point, and, if so, how we should proceed.

Before I end I should recall that each one of you have been invited here in your personal capacity. As you know in this Consultation you represent nobody but yourselves. The voice of governments we will hear later – when the FAO Committee of Fisheries will review and debate the report of this expert Consultation.

I will probably not be able to spend as much time with you as I would like to. But, I hope to be back at least on Friday afternoon in time for the adoption of your report.

Once more – thank you for taking time to come here and I wish you a successful Consultation.

DRAFT GUIDE FOR IDENTIFYING, ASSESSING AND REPORTING ON SUBSIDIES IN THE FISHERIES SECTOR

by

Lena Westlund
FAO Consultant

ABBREVIATIONS

AMS	Aggregate Measurement of Support
DSA	Daily Subsistence Allowance
GFT	Government Financial Transfers
GDP	Gross Domestic Product
GATT	General Agreement on Tariffs and Trade
IRR	Internal Rate of Return
ITQ	Individual Transferable Quotas
M	Million
OECD	Organization for Economic Co-operation and Development
PSE	Producer Subsidy Equivalent
R & D	Research and Development
SCM	Agreement on Subsidies and Countervailing Measures
TED	Turtle Excluder Device
US$	United States Dollars
VAT	Value Added Tax
WTO	World Trade Organization

TABLE OF CONTENTS

LIST OF APPENDIXES

LIST OF TABLES

LIST OF FIGURES

21

LIST OF BOXES

LIST OF BOXES

1 A FISHERIES SUBSIDIES GUIDE

This Guide has been developed to assist in studying fisheries subsidies. It aims at being an instrument for studies covering all different types of subsidies in all sectors of the fisheries industry and attempts to provide a flexible technical tool that can be used by those who prepare reports and studies on subsidies in the fisheries sector according to their specific needs.

As such, the Guide does not provide a rigid definition of subsidies but offers a framework for how to define fisheries subsidies . The Guide does not take any position with regard to whether a subsidy is "good" or "bad". Subsidies can be either positive, i.e., increasing profits of the industry, or negative, i.e., decreasing profits, but the analysis of the link between this impact and the eventual effect on resources and trade is beyond the scope of the Guide. Hence, the Guide does not cover the analysis of the effects of subsidies on resources, fisheries and trade but aims at assisting in collecting and organising the data on which these analyses could be based. This covers defining, classifying and quantifying fisheries subsidies as well as investigating the processes by which subsidies are provided.

Figure 1: Scope of the Guide

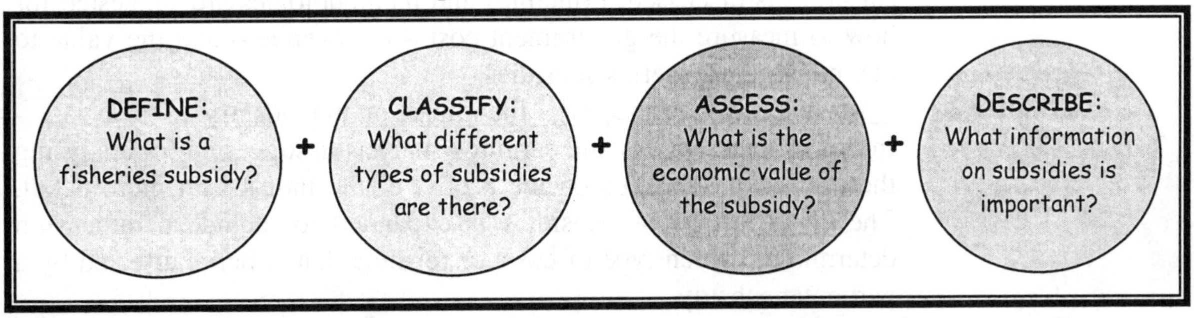

DEFINE: What is a fisheries subsidy? + CLASSIFY: What different types of subsidies are there? + ASSESS: What is the economic value of the subsidy? + DESCRIBE: What information on subsidies is important?

The Guide is based on the main principles agreed on in the FAO Expert Consultation on Economic Incentives and Responsible Fisheries, held in Rome on 28 November – 1 December 2000[1]. In early 2002, a preliminary draft Guide was prepared, based on available literature and information. This draft was thereafter tested by the carrying out of prototype studies in four different countries and it was then revised incorporating the experience from the test studies. The definitions and methodologies presented in the Guide have thus been developed by combining available theoretical knowledge with practical experience.

However, the subject of fisheries subsidies is vast and complex and the work carried out so far has not been sufficient for dealing with all aspects and issues that it includes. Therefore, this Guide should not be considered a final product but rather a flexible document that may need to be revised as more experience on how to study fisheries subsidies is acquired.

---

[1] See FAO 2000a.

2   HOW TO USE THE GUIDE

Following the two introductory chapters, the Guide is organized in eight main sections, each dealing with a separate aspect of a fisheries subsidies study:

Chapter 3   Planning and preparing for a fisheries subsidies study
Before starting the actual work on a fisheries subsidies study, there are several aspects to think about with regard to the objective, the scope and the resources needed.

Chapter 4   What is a fisheries subsidy?
This chapter discusses the definition of a fisheries subsidy. It emphasizes the importance of having a good understanding of both the fisheries sector and of the economy of which it is a part in order to be able to define subsidies.

Chapter 5   Different categories of subsidies
To facilitate the analysis of subsidies and to organize our information, it is suggested that fisheries subsidies are classified into four different categories.

Chapter 6   Assessing subsidies
This is the core of the Guide where the assessment of different types of subsidies is discussed. Principles and methodologies are suggested for how to measure the government cost – or revenue – and the value to the industry of fisheries subsidies.

Chapter 7   Costs and earnings analysis – the impact on the industry
In this chapter, a costs and earnings analysis is suggested for analysing the impact of subsidies on the profits of the industry in more detail. The classification of subsidies is expanded to include a dimension determining which type of costs or revenue that is being affected by a particular subsidy.

Chapter 8   Comparative analysis
Building on the results from the assessment of subsidies, this chapter suggests ways of examining the relative importance of fisheries subsidies.

Chapter 9   What to include in a subsidy description
In addition to assessing subsidies, such as estimating their economic value, there are other aspects that may be important to document and report. A checklist for what to include in a description of a fisheries subsidy, based on the information required for the WTO subsidy notification procedures, is suggested.

Chapter 10   Reporting on subsidies
In this last chapter, a few ideas are given with regard to how to structure the report on a fisheries subsidies study.

The user of the Guide should chose those parts of the document that are relevant for a particular study. Not all sections may be of interest to everyone. It is however recommended to read through the whole Guide to start with to gain a good understanding of the subject field and the issues at stake. Even though the Guide makes an effort to present the topics in the same order as they are likely to occur in the work on a fisheries study, there are several cross-cutting issues that are not possible to present in the assumed sequential order. It should also be recognized that the work is likely to be an iterative process that will take us backwards and forwards in the Guide several times. The **index** at the end of the document may be useful in this respect.

To illustrate various approaches and methodologies, several examples have been included in the Guide – in particular on how to asses subsidies – and are presented in "boxes" throughout the text. These examples refer to an invented country called Seidisbus and are not real subsidies as such. However, the contents of the examples are in most cases derived from the experience of the four prototype studies carried out for the preparation of this document. Hence, it is believed that the examples cover some of the more important subsidies as well as the most common problems related to their assessment. At the same time, it should be pointed out that it has not been possible to include all aspects and types of subsidies in the document and the Guide does not lay claim on being exhaustive or fitting all situations equally well; omissions are inevitable. It is still hoped, though, that the guidelines will help identifying and assessing all types of fisheries subsidies, whether mentioned explicitly in the Guide or not.

A **bibliography** is included at the end of the report and a glossary is found in **Appendix I**, giving explanations to some of the economic expressions and terms used in the Guide.

3 PLANNING AND PREPARING FOR A FISHERIES SUBSIDIES STUDY

As with all research projects and studies, there is a fair amount to do before the actual survey work can start. Planning for a study is an important process and the quality of the preparations can be essential for how good results we achieve. Hence, sufficient time and effort have to be allocated for the planning phase of the work. Some of the questions we need to answer before we start a fisheries subsidies study are:

- What is the objective of the study, i.e. what questions is the study trying to answer?
- What scope should the study have? Should it cover the whole of the fisheries sector or only selected subsectors or regions? Should it cover all subsidy types or only certain categories?
- What are the resources available for the study? How much time do we have?
- Who should carry out the study? What competences need to be represented on the study team?
- What other preparations are needed with regard to background reading and development of methodologies?

The clearer the **objective** of the study, the more focused the work can be and better results are likely to be achieved with fewer resources. If the study is carried out in a given context, maybe requested by a department or ministry with defined terms of reference, the objective is likely to be defined as well. However, if this is not the case and if it is the first time a review is to be carried out, we may want to keep the study quite broad, giving more of a general inventory of existing subsidies and related issues. A first subsidy study may have as an objective to identify issues to be studied further.

The **scope** of the study is in some ways closely related to the next question in the list above, i.e. what **resources are available**. The more extensive the study, the more resources and time we will need. Time and resources needed will also depend on the general availability of data and the size and complexity of the fisheries sector. It is of course easier to carry out a fisheries subsidies study in a small country with a relatively limited fisheries sector for which there is already a good data collection system than in a big country with a very large – and maybe dynamic – fisheries industry for which data are generally not available with the central administration.

Nonetheless, the minimum time required for a study also in a relatively "easy" country or region is probably at least three to six months if we want to cover the whole fisheries sector, i.e. all subsectors including input industry, capture fisheries, aquaculture, processing, and marketing and distribution. This time is likely to allow us to identify the main existing subsidies and to assess and give values to most of the more direct subsidies of categories 1 and 2 (see chapter 5). Of course, the time required also depends on the number of people in our study team and on whether we work full-time on the study or not. However, even with a large team and full-time engagements, there are going to be delays in the data collection considering the large number of contacts that will be needed for a complete sector coverage.

If we want to examine some of the more complicated subsidies in more detail, i.e. categories 3 and 4 subsidies covering longer-term effects and non-interventions (see chapter 5), considerably more time will be needed. To examine issues such as free access to resources or the actual effects of gear regulations may require a focused effort. In fact, we may want to organize the study in a way that allows a separate study team to look into details of specific issues. If only a limited number of these issues are being included and the special efforts organized at an early stage – to be carried out in parallel with other tasks – time can be saved. Nevertheless, it would be expected that a more detailed fisheries subsidies study would require at least six months to a year to complete, from the starting of the planning phase to the finalization of the report.

In addition to ensuring that we have enough time, it is very important that our **study team** includes the right competences. With team members that are already familiar with the fisheries sector, the economic framework of the country or region, and with the concept of subsidies in general, the work on the study is likely to be easier. However, we should be more specific than that and, firstly, it would appear absolutely essential that at least one person – the team leader – has excellent analytical skills in addition to a general knowledge of the sector and the issues as well as the necessary time and interest for investing in the study. It is important to point out that the Guide is just that: a guide for facilitating the conduct of a fisheries subsidies study. It does not, however, give a complete set of "rules" to follow or set forms to fill in. It provides a basis for creating the tools needed to carry out a study and guidelines for the identification, classification, description and assessment of subsidies in a systematic way. Still, someone needs to develop these tools and analyse the results.

Much of the work on assessing subsidies, in particular with regard to government costs, is based on information from the public accounts. Many of the calculations use accounting principles and standard methods for financial analysis. It would hence also appear important to have at least someone on the study team who is an accountant who is familiar with the structure of the government accounting system. The team should also include an economist. Without these two disciplines, misunderstandings could occur and the results become unreliable. Likewise, access to information is fundamental for the study and this aspect also needs to be taken into consideration when appointing our study team members.

With regard to **other preparations** before starting the actual survey, it is usually always a good idea to conduct general interviews with key informants and resource persons as well as to consult existing relevant documentation in order to establish a preliminary overview of the structure of the economy, the fisheries sector and existing subsidies. This preliminary investigation is particularly important for study team members who are not familiar with the subject area.

Once we have a plan for what we would like to do, it is recommended that we put it down in writing. Clear terms of reference and a timeframe for the work, including milestones for the completion of various subcomponents, will help us to keep the right focus of the study and to monitor the work.

Figure 2: Planning and preparing for a fisheries subsidies study

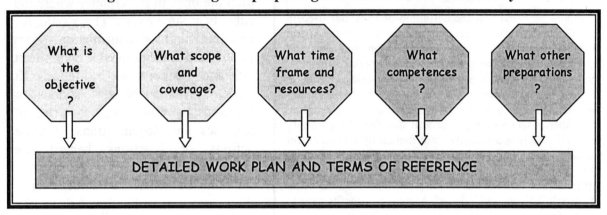

4   WHAT IS A FISHERIES SUBSIDY?

4.1   A definition of fisheries subsidies

After having thought through the overall organization of the work and drawn up a general plan for the task ahead of us, the next thing we have to do when starting a fisheries subsidies study is to define what we mean by a fisheries subsidy.

To most of us, a subsidy is some kind of government support – mostly of a monetary nature – to the private sector, generally serving a public purpose. Looking up the term "subsidy" in a dictionary gives us that a subsidy is "a direct or indirect payment, economic concession, or privilege granted by a government to private firms, households, or other governmental units in order to promote a public objective" (FAO Fisheries Glossary and Encyclopaedia Britannica 2001).

Based on this general notion, the Guide proposes a broad definition of fisheries subsidies according to which a subsidy could fundamentally be any government intervention – or lack of intervention – that affects the fisheries industry and that has an economic value. This economic value is interpreted as something having an impact on the profitability of the fisheries industry. The intra-government aspect from the definition above is disregarded and subsidies are defined as actions or inactions for which the recipient is part of the private fisheries industry (and not "other government units").

However, not everything the public sector does or does not do can be classified as subsidies[2] and a further qualification of the definition is needed with regard to reference points.

---

[2] For clarity, it should be mentioned that the Guide generally uses the term "subsidies", meaning – as defined above – all government actions and inactions. Occasionally, the term support measure or program is used as a synonym. With regard to government costs (revenues) for subsidies, expressions like public expenditures,

Accordingly, a subsidy should be something that is out of the ordinary, i.e. something that is done – or not done – outside of normal practices:

Fisheries subsidies are government actions or inactions that are specific to the fisheries industry and that modifies – by increasing or decreasing – the potential profits by the industry in the short-, medium- or long-term

---

Box 1: WTO definition

It should be noted that the definition of subsidies used in this Guide is much broader than the one used in the World Trade Organization (WTO) Agreement on Subsidies and Countervailing Measures (SCM) which is perhaps the most commonly cited and practically applied subsidy definition. The SCM Agreement is WTO's basic subsidy agreement and the one that currently governs trade disputes regarding the fisheries sector in this respect. It specifies that a subsidy exists if "there is a financial contribution by a government or any public body within the territory of a Member" and this contribution fulfils certain specified conditions, or if "there is any form of income or price support in the sense of Article XVI of GATT 1994". Moreover, benefits have to be conferred. For the subsidy to be offending, it also has to be "specific", "prohibited" or "actionable" and cause "adverse effect" (WTO 1994 Agreement on Subsidies and Countervailing Measures, article 1, also described in Milazzo 1998).

"*Government*" here also includes other governments and public bodies than the ones in the country where the subsidy as such exists. This would, for example, include contributions from public and international development aid and cooperation institutions. It also of course includes actions or inactions by non-fishery government agencies and organizations. If these actions or inactions benefit the fisheries industry in a significant way, they may be fisheries subsidies even if they are not only directed to the sector. Sponsorships by private companies do however not constitute subsidies.

The "*fisheries industry*" refers to all productive subsectors of the fisheries and aquaculture sector, i.e. all types of input industry – including transport and other support services – capture fisheries, aquaculture, processing and marketing. It covers all producers and operators, both small and large-scale, engaged in recreational, subsistence and commercial activities. For our particular study, we may of course have decided that we only want to look at one or a few subsectors (see chapter 3).

By "*potential profits*" the overall profitability of the industry is implied. While subsidies affect profits in the short, medium and long-term, the Guide's focus is on the more direct shorter-term financial effects as we will see when discussing how to assess subsidies in chapter 6. It should be noted that subsidies can also negative, i.e. decreasing profits. Examples of negative subsidies would be taxes and other fees and duties.

4.2   The context

When we say that a fisheries subsidy is an action or inaction that is specific to the fisheries sector, we need to know what we mean by "specific" in order to be able to distinguish between subsidies and non-specific – or general – actions and inactions. The best way to do this is to define what is general and use this situation as the benchmark for "normality" against which specificity is measured, i.e. anything that is different from our normality reference point is specific and hence a subsidy.

---

expenses or costs, and public budget implications are used meaning the same thing unless stated otherwise. Likewise, the fisheries industry is sometimes called the private sector or referred to as fisheries operators, companies or firms.

Figure 3: Definition of fisheries subsidies

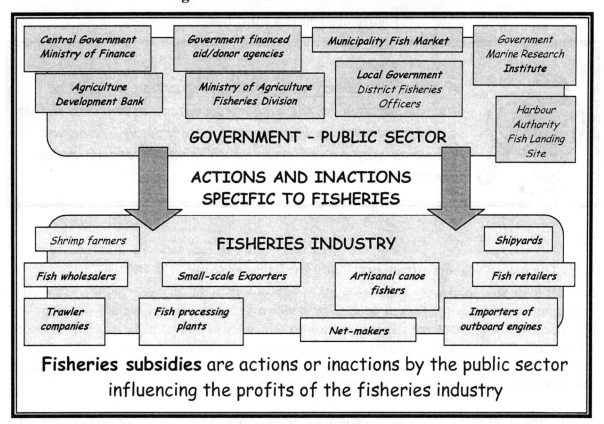

It is conceivable that sometime in the future countries agree on what is "normal" and therefore useful as benchmarks. Such benchmarks could *inter alia* include: interest rates for investment loans, standard fuel prices, and minimum levels of cost recovery for fisheries management, etc. However, because such standards do not exist and because it will take time to reach agreement about them the Guide suggests - for now - using standards represented by the overall economic framework of the particular country – or region – under study[3]. The reference points to be used should refer to other sectors in the country, or group of countries, i.e. the "normal" situation without subsides is represented by the circumstances industry in general operates under in the country or region and fisheries subsidies are defined and measured as deviations from these conditions. For example, in a country where public services are provided so to say free of user charge – because they are financed through the tax system – it would be considered normal that also the fisheries industry benefits from certain services without them being defined as subsidies. On the other hand, in a country where cost recovery is the norm, the same benefits to the fisheries sector are subsidies if not directly paid for by the industry.

It is hence up to the user of this guide to define what should be considered subsidies in his/her particular context, based on the specificity of the economy concerned and its policy framework.

---

[3] On those occasions when the results of the subsidy study are to be used in an investigation of the effects of subsidies on international trade in fish and fish products, the analysts will need to adapt the study according to the definition of subsidies that is contained in the WTO agreement on Subsidies and Countervailing Measures.

Figure 4: Fisheries subsidies

A fisheries subsidy is a government action or inaction that is specific to the fisheries sector, as defined within a given economic policy context. Accordingly, fisheries subsidies may be different in different economic policy contexts.

4.3 The macro-economic framework and the fisheries sector

In order to be able to identify fisheries subsidies according to the above definition, we need a good understanding of the overall economic framework in the country we are studying. We also need to know about the fisheries sector so that we can compare the conditions under which the fisheries industry operates with the situation of the industry in general. Quantitative information on several economic and financial issues is also needed for assessing the value of the subsidies, something that is discussed in chapters 6 and 7 below.

A **baseline survey of the macro-economic framework** to be used for identifying and assessing fisheries subsidies could include the following aspects:

- *Policy and legal framework*
  We need to understand the overall political framework with regard to the role of the public sector. What is the general policy with regard to cost recovery of public services? What government market interventions are there? What is the role of state-owned enterprises? Are there support programmes for businesses, e.g. for regional development, infant industry, etc.?

  We should also look into more specific policy and legal matters, particularly in areas that are likely to be relevant to the fisheries sector, e.g. environmental policies and general pollution control measures, food safety and sanitation standards, labour and employment laws, and practices with regard to user right policies in the natural resources sector.

- *Interest rates and monetary policies*
  Information on commercial interest rates is needed for determining whether the fisheries industry benefits from loans on favourable terms and for assessing other possible subsidies related to investments. As far as possible, the real alternative financial cost should be used as the benchmark and where different commercial interest rates are used for different types of investment – usually according to the perception of risk – care should be taken to use the rate corresponding to the particular situation. The baseline survey may hence need to list different commercial rates. It should be noted that this might mean including usury rates in an informal market if this is the factual market alternative. If information on commercial rates is lacking, an appropriately adjusted inter-bank offered rate or the rate of return for government bonds might serve as useful guidance.

Information on other general loan and credit conditions, such as the need for collateral or typical amortization periods, could also be included in the baseline survey, as well as information on monetary policies of importance with regard to the credit system.

- *Tax regulations and social security coverage*
  Information on the overall public revenue and tax system that might be relevant includes VAT rates, for different goods and services if differentiated, and corporate tax rates including general rules for, among other things, depreciation allowances for fixed assets and tax deference possibilities.

  With regard to employment and labour costs, the main features of the tax systems for individuals need to be covered as well as the general social security structure, including rules for obligatory contributions and common benefits.

- *Trade regulations and border measures*
  Tariff and non-tariff border measures should be looked into, giving details on differences between different types of goods and services. Border restrictions with regard to direct investments, ownership or the exertion of certain professions or commercial activities also need to be considered as well as regulations with regard to immigration and labour movements.

- *Exchange rates and currency regulations*
  We need to know if there are any capital movement or foreign currency restrictions, especially if they apply in different ways to different sectors of the economy. Official and unofficial exchange rates of importance to the business community should also be looked at.

Box 2: Choosing benchmarks

Sometimes it is difficult to define the norm value and there may be several alternative benchmarks. For example, there may be a wide range of customs tariffs for different types of products making it virtually impossible to decide what is the "normal" level. In this case, we may decide to compare the fisheries sector with one or a few other major economic sectors that are similar to the fisheries sector. With regard to the example of customs tariffs, the norm value used for fish and fishery products could be the tariff level applicable to food and agricultural commodities. Another example concerns taxes for which the situation of small independent artisanal fishers, aquaculturists and traders/processors could be compared with small-scale farmers while larger fishing, aquaculture and processing companies are maybe better compared with light manufacturing industry and other food processing industry. In many (developing) countries, the former belongs to the informal sector whilst the larger operations are part of a more organized business structure, operating under quite different conditions (for example with regard to VAT registration and refund).

In the next step on the way to identifying fisheries subsidies, we compare the fisheries sector against the background of the overall macro-economic framework and the benchmarks we have defined. Accordingly, we need to investigate if the fisheries sector is treated differently from other economic sectors or if the economic and political conditions that the fisheries industry operate under are different from those of other industries. In a **review of the fisheries sector**, we would like to identify and investigate all <u>fisheries specific policies and regulations</u>, e.g.:

- Public market interventions in the fisheries sector, e.g. price policies, state-owned enterprises, support programmes, etc.
- Fisheries management and access right regimes
- Water and land use regulations for aquaculture
- Fish safety and hygiene regulations
- Capital markets and credit schemes relevant to the fisheries industry
- Tax regulations for the fisheries industry
- Social security policies and regulations with regard to labour in the fisheries industry
- Trade regulations and border measures regarding fish, fishery products and input goods

4.4 Where and how to find the information

The bullet points in chapter 4.3 above cover a large number of different types of information needed for the fisheries subsidies study. Depending on the competences and positions of the people in the team carrying out the study, some of this information may be easy to find, e.g. in a ministry or department of fisheries, information on fisheries management regimes is likely to be readily available. General information on issues that we already know something about, for example, regarding economic policies or tax regulations, may also be quite straightforward to get. In the same way, some subsidies may be easy to identify. Certain support schemes, e.g. an investment grant scheme, may be commonly known and therefore not difficult to investigate.

.

However, there are other issues that may be outside our normal work area or field of competence. There may even be aspects that we do not know about at all and consequently a risk that certain subsidies are not discovered. Therefore, we need to consult a wide range of information sources. To ensure to the extent possible that all subsidies are identified, it is suggested that we approach the exploration of subsidies from two angles, i.e. from the point of view of the provider and from the point of view of the recipients.

In order to do so, we need to identify:

- *All government agencies and organizations that are involved in the fisheries sector and that may provide subsides – in one form or another – to the sector.*
  The relevant bodies may be fisheries specific agencies – such as a fisheries ministry or department – or non-fisheries, e.g. a research institute or a health authority. There may also be international organizations, or foreign donor governments. Semi-public and interest groups in the form of, for example, trade unions or producer organizations, should also be identified. They are likely to provide support to the industry that they in their turn receive public funding for.

- *The actors in the fisheries industry who receive or benefit from fisheries subsidies.*
  These are the companies and individual operators that are active in fishing, aquaculture, production of inputs, processing, marketing and trade, etc. Making an inventory of the economic activities of the sector and listing main groups of companies and operators according to subsector, size, business form, etc., will give a good picture of the organization of the fisheries industry and help us identify who should be consulted in the process of identifying and assessing fisheries subsidies.

Accordingly, it is recommended that we carry out an in-depth institutional survey as well as a thorough review of the fisheries industry. The latter is further discussed in chapter 7.

Figure 5: A two-way approach

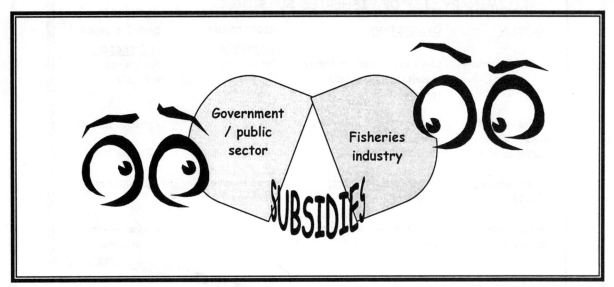

4.5   Identifying fisheries subsidies – a first list

Once we have identified all the stakeholders i.e. all parties involved or concerned with the fisheries sector, and we also have a fairly clear idea of what information we need, we can start collecting information and identifying the relevant subsidies. The information gathering can take many different forms – review of existing documents and reports, questionnaires, interviews, discussions, telephone calls, enquires by letter, etc. – and methodologies suitable for each particular situation have to be adopted. We may occasionally also want to consult people outside our group of stakeholders, i.e. companies or representatives of professionals in economic sectors outside fisheries, to clarify benchmarks and reference points.

The extent of the data collection and the depth of analysis will depend on the objectives of our particular study and the terms of reference that we have given it. It is likely that we want to investigate far beyond a simple listing of existing fisheries subsidies and the subsequent chapters of this Guide suggest and discuss various types of analyses. However, whatever the level of detail we have agreed on for our study, it is generally a good idea to draw up a list early on of already identified subsidies and of situations and support measures that could potentially be defined as subsidies once we have investigated them further. This working document could also contain short descriptions of the subsidies, and information on the responsible authority (provider) and the recipients, and will – when finalized – provide a summary overview of the subsidies in the fisheries sector.

Figure 6: Example of a preliminary list of fisheries subsidies

PRELIMINARY LIST OF FISHERIES SUBSIDIES

Subsidy	Description	Responsible authority	Recipients / beneficiaries (subsector)
1. Fuel tax rebate	Lower tax on fuel for fishing vessels (reimbursement retroactively).	Department of Fisheries	Fishers registered with DOF
2. Provision of landing site facilities	Landing sites provided free of charge	City municipalities	Artisanal subsector
3. Investment grants	Storage and transport equipment	Department of Fisheries	Aquaculturists
4. No resource access fees	Fishing vessels do not pay licence fees	?	Marine capture fisheries subsector

5 DIFFERENT CATEGORIES OF SUBSIDIES

5.1 Four categories of fisheries subsidies

When working on identifying subsidies, we will realize that there are many different types of fisheries subsidies. Some situations and measures can quite indisputably be identified as subsidies, e.g. grants and other direct financial transfers from the government to the private fishing industry, while we may have more doubts with regard to a situation of, for example, apparent lack of pollution control. Moreover, depending on the objective of our study, we may have decided that we only want to look at the most direct subsidies and that we do not have any interest or time for carrying out the often quite comprehensive and time-consuming analyses related to more indirect or non-intervention subsidies (see also chapter 3).

Consequently, to facilitate the organization and analysis of our subsidies information, the Guide suggests that we classify fisheries subsidies into four main categories, i.e.:
1. Direct financial transfers
2. Services and indirect financial transfers
3. Interventions with different short and long-term effects
4. Lack of intervention

5.2 Category 1: Direct financial transfers

The first category includes **all direct payments by the government to the fisheries industry**. These subsidies have a direct short-term effect on the profitability of the industry and can also be negative. Their cost (revenue) to the government can usually be found in the public budget and its direct value to the industry will appear directly in the cash flow of the recipient industry. Subsidies belonging to this category are easy to identify and it would generally not be difficult to find consensus when defining these subsidies.

Examples of Category 1 subsidies include:

Investment grants (e.g. to purchase vessels or for modernization), grants for safety equipment, vessel decommissioning programmes, equity infusions, income guarantee schemes, disaster relief payments, price support, direct export incentives, etc.

Negative subsidies in this category would include, for example, various taxes and fees, and import/export duties.

5.3 Category 2: Services and indirect financial transfers

The second category covers **any other active and explicit government intervention but which does not involve a direct financial transfer as specified under Category 1**. Category 2 subsidies also have a direct short-term effect on profitability but are rarely negative. Their cost may or may not be specified in the public budget and the value to the industry does usually not appear explicitly in the accounting of the recipient industry. Many of the subsidies in this category are services of some kind provided by the public sector or indirect financial transfers.

There are four sub-groups in this category which are listed below together with examples from each group:

- *Category 2A: Non-tariff border measures and other market interventions, e.g.:*
  Import quotas, export promotion support, direct foreign investment restrictions, etc.
- *Category 2B: Tax and duty exemptions and other reduced charges by government agencies, e.g.:*
  Fuel tax exemptions, investment tax credits, deferred tax programmes, special income tax deductions, etc.
- *Category 2C: Services provided by the government that are generally also provided by the private sector but under different – less beneficial – conditions, e.g.:*
  Investment loans on favourable terms, loan guarantees, special insurance schemes for vessel and gear, provision of bait services, etc.
- *Category 2D: Services provided by the government that are generally not provided by the private sector and for which the full cost is not recovered, e.g.:*
  Inspection and certification for exports, specialized training, extension, ports and landing site facilities, payments to foreign governments to secure access to fishing grounds, government funded research and development programmes, fisheries management, international cooperation and negotiations, etc.

5.4 Category 3: Interventions with different short and long-term effects

Our third category of fisheries subsidies allows us to consider a longer time perspective and includes **government interventions that have a negative economic impact on the industry in the short-term but ultimately result in long-term benefits** (with regard to, for example, the resource base) and/or more general benefits to society as a whole (with regard to, for example, the environment). The cost of Category 3 subsidies – usually an administrative cost

– may be accounted for among other public expenditures for management and regulations and difficult to identify. The short-term value to the industry would commonly appear as an expenditure in the accounting of the industry while the positive long-term effects are implicit.

Some examples of Category 3 subsidies are:

Environmental protection programmes, gear regulations (e.g. Turtle Excluder Devices), chemicals and drugs regulations, etc.

5.5 Category 4: Lack of intervention

The last and fourth category covers the area of **lack of government intervention** and may be the most difficult one to deal with. Category 4 comprises inaction on behalf of the government that allows producers to impose – in the short or long-term – certain costs of production on others, including on the environment and natural resources, and that has short-term positive effects on the industry's revenues and/or costs. These subsidies are usually positive in the short-term but negative in the long-term. By definition, they do not imply a cost to the government and their value to the industry is implicit.

Examples of this type of subsidies include:

Free access to fishing grounds, lack of pollution control, lack of management measures, non implementation of existing regulations, etc.

Box 3: What to do when there is more than one suitable subsidy category?

Some subsidies may fall into several different categories at the same time. Duty on imports of fish and fishery products, for example, may protect local producers and would hence be a positive Category 2 subsidy for the processing industry. At the same time, it may be a negative Category 1 subsidy for fish importers and retailers who are paying the import duties. Likewise, most subsidies are impacting not only on the direct recipient of the subsidy but will be "carried along" in the downstream production and distribution chain and it could at times be difficult to determine which stage in the chain should be considered the true direct beneficiary. Market facilities at a landing site, for example, benefit both the fishers – the sellers – and the traders and processors – the buyers. When studying subsidies, each measure and support program has to be analysed individually and should be classified according to its particular characteristics. Naturally, there are subsidies that are difficult to put in one "box". The criteria for classification should be the most direct and main impact on revenues or costs. There are often second-stage and indirect consequences as well as side effects but, even though their importance is recognized, these should not be the main concern for the classification. The general rule should be to only classify subsidies in one category but there may of course be occasions when this is not feasible or unpractical. It has to be recognized, though, that sometimes we have to make assumptions and simplify our analysis for practical reasons.

It should also be remembered that the examples of subsidies and of how to classify them in this Guide are only there for guidance. We may find in our particular fisheries subsidies study that we want to classify certain measures or situations in a different way and we will most probably find subsidies that are not specifically mentioned in the Guide.

Figure 7: The Guide's four subsidy categories

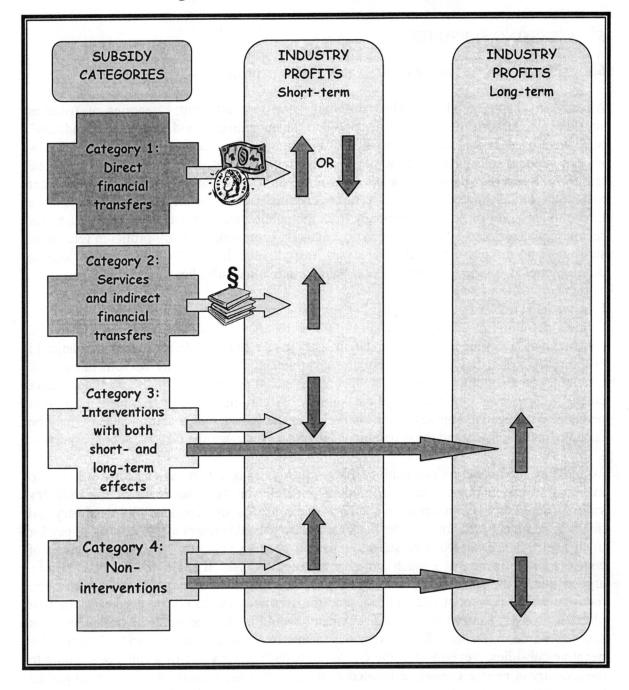

As can be seen, this classification in categories builds on the modality of the subsidy, i.e. whether it is based on a direct financial transfer or not (Categories 1 and 2), how the subsidy modifies industry profits in the short- or medium/long-term (Category 3) and whether it involves an action on behalf of the government or not (Category 4). Categories 1 and 2 – with the exception of some of the government specific services such as fisheries management – correspond quite closely to other definitions used in practice by, for example, WTO, while Categories 3 and 4 include more implicit benefits (or detriments) to the industry.

There are of course many ways of classifying subsidies and also many possible subcategories available. Some of the main aspects found in the literature according to which subsidies can be classified are reviewed in **Appendix II.** A list containing more examples of subsidies from the different categories presented here is found in **Appendix III**.

6   ASSESSING SUBSIDIES

6.1   Government cost and industry value – some principles

Already in chapter 5 above, when discussing the four different categories of fisheries subsidies, the notions of "cost – or revenue – to the government" and "value to the industry" were mentioned. In this section, we will look closer at these concepts and try to establish how we can measure subsidies, i.e. how we can assign monetary values to fisheries subsidies. The reason for wanting to do so is to allow us to make more quantitative analyses. We may want to be able to say how much of the profits generated by the fisheries industry that can be linked to subsidies or how much money from the public budget that is spent on the fisheries sector. Chapters 7 and 8 below give more details on examining industry profits in relation to subsidies and discuss comparative analyses, respectively. In this chapter, we will look at different methodologies and practices for valuing individual subsidies of different types.

First we need to clarify the concepts of <u>government cost (or revenue)</u> and <u>industry value</u>. The Guide's definition of subsidies is centred on the impact a subsidy has on firms' profits and the value of a certain subsidy for the industry could thus be defined as this change in profitability that it has provoked. For the government, the value may be something quite different, corresponding to the public expenditure including direct and indirect administrative costs for operating and managing the subsidy. Accordingly, the government cost – or revenue, if a negative subsidy – and the value to the industry of a certain subsidy may be different and to understand the economic value of a subsidy, we should look at both aspects.

To measure the impact on profitability that a subsidy has is not an easy task and we need to make some assumptions. The main guiding principle for the **assessment of the industry value** is based on the assumption that all positive subsidies are beneficial to the industry and that if governments did not provide them, the industry would either have to or want to pay for them itself. Hence, all positive subsidies have a positive value to the industry. This is of course a simplification of a much more complicated analysis. It could, for example, also be argued that if the support provided through subsidies was necessary for the industry, the industry would pay for it itself already and the subsidies would not be needed in the first place. Moreover, somewhat different arguments would need to be used for certain subsides such as border measures, fisheries management, free resource access and of course for negative subsidies. Nevertheless, the Guide suggests that the value of a subsidy accruing to the industry is most accurately estimated as the cost that the industry would have to pay on commercial terms for obtaining the same service or good and that this principle should be utilized whenever possible.

With regard to the value of fisheries subsidies expressed as the **cost (revenue) to the government**, the assessment has to be based mainly on information from the public budget, except for in the case of foregone revenues (e.g. tax rebates) which are generally not included in the budget and will have to be assessed separately. Moreover, it is important that the cost of a subsidy is evaluated not only as the financial transfer it may entail – in the case of a grant for example – but that also the administrative cost to run the program or implement the regulation is included. This implementation cost includes personnel cost and other operational costs incurred by, for instance, fisheries administrations or other government agencies dealing with subsidies reaching the fisheries sector[4].

For some types of subsidies, in particular those belonging to Category 1 and consisting of direct financial transfers, assessing the value according to these principles may be a relatively easy task. It is probably fair to assume that the approximate value to the industry of an investment grant corresponds to the amount of the grant plus the interest rate it would have cost to borrow the same amount of money on commercial terms. For the government, the expenditure would consist of the grant itself and the administrative cost related to its distribution. Other subsidies are more difficult to assess, in particular with regard to deciding the value the subsidy represents to the industry. The value is the change in profits but this change should be measured compared to what? Disregarding various economic theory considerations[5] and taking a practical approach, it is suggested that the change is measured as the difference between a situation *with* the particular subsidy and a situation *without* it. This approach is also in line with how we identify fisheries subsidies, i.e. by the definition of benchmarks representing the "normal" – non-subsidy – situation. Accordingly, in cases when the subsidy means offering services or goods at a price different from an existing market price, such as favourable loans at low interest rates or a decommissioning program paying for vessels to be scrapped, the real benefit to the industry consists of the difference between the price offered through the subsidy measure and the market price the industry would have paid or received for the service or good had the subsidy not been available.

For subsidies for which there exist commercial alternatives at market prices, the calculation of the value of the subsidy is hence relatively apparent. Also subsidies such as tax rebates that clearly represent a situation where the fisheries industry is treated differently from industry in general are fairly uncomplicated to assess. The value to the industry would in such a case be the difference between the tax actually paid and the tax that would have been paid had there not been a subsidy. Here the benchmark is not the market price – because there is no market price for taxes – but the normal tax rate. The cost to the government should be calculated as the revenues foregone owing to the scheme plus any additional cost involved in administrating the scheme.

As already noted, the situation becomes much more complicated when we look at areas such as fisheries management and resource access, typically some of the services included in Category 2 as well as Categories 3 and 4 subsidies that include longer-term effects or government inaction. For some of the former, i.e. services provided by the public sector in Category 2, we may know the cost to the government but what is the value of the service to the industry? Here the principle of market prices and norm values cannot be used, simply because there are none. In these situations, we may want to assess the actual impact on the profits of the industry, generally in the form of increased revenues. For example, if we can

---

[4] This is of course an oversimplification of reality and it could be argued that to calculate the real cost to the public sector and society, a much broader analysis would be needed looking at opportunity costs, externalities and long-term environmental impact. Some of these aspects may be captured in the assessment of the value of the various subsidies but the fact remains that a reliable in-depth welfare economics evaluation of the value of fisheries subsidies is difficult in practice. Hence, a quantitative analysis in this respect is considered beyond the scope of the type of study suggested by the Guide but any qualitative information that can be added to the study results would of course be valuable.

[5] Schrank and Keithley Jr. (1999) discuss the possibility of using the theoretical concept of *perfect competition* as the benchmark when assessing subsidies. However, they conclude that the criterion is "too amorphous to serve as an operational concept" (Schrank and Keithley Jr. 1999, page 157). Another suggested benchmark is the situation of *efficient allocation* of resources (OECD 1993), discussed in Box 4. This is closer to the practical definition recommended by the Guide but the theoretical criterion cannot be fully fulfilled and the concept is thus not used.

estimate the value of increased sales of fishery products owing to an "Eat Healthy Food" and fish consumption campaign, the value of this campaign to the industry would be the increase in net income thanks to the improved sales.

However, this type of calculation is often a cumbersome exercise requiring a substantial share of research that is not always a practical option for our fisheries subsidies study. Hence, in these cases we may need a proxy for our estimate and the Guide suggests using the cost to the government – if known or estimated – also as an estimate of the value to the industry.

Box 4: Theory and practice

In the agriculture sector, various methods have been developed for measuring subsidies in relation to trade distortions. One tool commonly used is the Producer Subsidy Equivalent (PSE) which has been the basic Aggregate Measurement of Support (AMS) in GATT/WTO trade negotiations. According to the OECD, "the PSE is an indicator of the value of the transfers from domestic consumers and taxpayers to producers resulting from a given set of agricultural policies, at a point in time" (Silvis and van der Hamsvoort 1996, page 529). It would naturally be interesting to use this type of measurement also for the fisheries sector. However, work by OECD (1993) has shown that due to the characteristics of the fisheries industry, the PSE is not a valid indicator for fisheries subsidies. Generally, in other sectors, the benchmark against which assistance and related trade distortions are measured is a situation of no government intervention, which corresponds to a situation of economically efficient allocation. In the fisheries sector, this benchmark situation of efficient allocation is much more difficult to assess as market failure is inherent to an open access fishery, implying that efficient allocation will only occur in a situation with government intervention. Moreover, it is difficult to establish external reference prices as well as domestic prices for raw fish – parameters required for the PSE model – because fresh fish is highly perishable and heterogeneous, and as a high degree of vertical integration is often found in the sector (OECD 1993). In addition, the restrictions to access by foreign vessels to domestic fishing grounds and to port facilities constitute a common public support in the fisheries sector that the PSE model does not take into account (Munk and Motzfeldt 1993).

No other single assessment methodology has been found and instead a variety of approaches is being suggested by the Guide for assessing the different types of subsidies. When identifying these approaches, attention has been given – to the extent possible – to their practicality, i.e. the methodologies recommended should be workable and give results easy to understand and verify. Hence, for example, shadow prices and opportunity costs have generally been excluded and it is the most direct effect that is measured, largely ignoring second-stage consequences. This approach could be considered unsatisfactory from a theoretical point of view but it is believed necessary for practical reasons.

There are unfortunately also cases when there is no easily estimated government cost to use when we have problems estimating the value to the industry. As we have seen in chapter 5.5 above, Category 4 subsidies do not incur costs to the government because they are non-interventions[6]. In these situations, we may have to turn to the use of standard or conventional values in order to quantify the impact of the subsidy on industry profits. An example of this type of situation, which is discussed further below, is free access to resources. If the industry is allowed to fish without paying for this access to a natural resource – or it is paying a fee that is considered below the actual value of the resource access – we may want to define this privilege as a subsidy and cost it in proportion to the value of the catches, using an estimated standard rate.

---

[6] See footnote 4.

Before moving on to looking at examples of how to value different types of subsidies in the next chapter, the main principles for how to assess fisheries subsidies are summarized in Figure 8.

Figure 8: Summary of main principles for assessing fisheries subsidies

ESTIMATES OF THE VALUE TO THE INDUSTRY TO BE BASED ON:

1. The corresponding market price value, if available, of the service or good provided through the subsidy. For investments, the financial cost (i.e. corresponding to commercial interest rates) should be included.

2. The normal rate or situation applicable to other industries and to the economy in general (e.g. tax rates, acceptable pollution levels, etc.).

3. An estimate of the net income effect of the subsidy, if there is no applicable market price or norm value.

4. The public cost, if there is no applicable market price or norm value, and an estimate of the net income effect is not possible.

5. A standard, conventional, value related to, for example, turnover, if no other value is available.

ESTIMATES OF THE COST (REVENUE) TO THE GOVERNMENT TO BE BASED ON:

1. The actual budgetary expenditure, when available.

2. Foregone revenues, when applicable (e.g. tax rebates).

3. Related administrative costs including personnel costs and a proportionate share of overhead costs.

6.2 Assessing different types of subsidies

6.2.1 Presentation of subsidy examples

Having looked at some general principles for how to assess fisheries subsidies, we will now discuss different types of subsidies in more detail. The text is organized in bullet points according to the main groups of subsidies that we are likely to come across and approximately following the structure of chapter 6.1, i.e. starting with some of the most direct Category 1 subsidies. Additional information on particular methodological issues and examples – using an invented country called Seidisbus – are given in boxes. These examples are summarized in Figure 9 at the end of the chapter.

6.2.2 Investment grants

An investment grant program is probably one of the most obvious examples of a direct financial transfer subsidy of **Category 1**. These schemes are commonly used for the purchase or modernization of equipment and facilities, having improved competitiveness through more efficient production as an objective. They generally benefit investments in fishing vessels or in the processing industry but can also be found in other parts of the sector, for example, in the input industry for shipbuilding or in the aquaculture subsector.

The value to the industry of this type of subsidy scheme consists of the value of the grant itself plus an estimate of the interest it would have cost to finance the investment commercially. Generally the investments are in fixed assets for which the cost is allocated over a depreciation period of several years

---

Box 5: Investment grants - An example

In the country Seidisbus, the Department of Fisheries operates a scheme through which aquaculture producers can apply for grants for improving their fresh fish storage and transport facilities. In the year 2000, a total of 25 aquaculture firms applied for and received funds for investments in cold storage and insulated trucks for a total amount of US$ 700 000. This was a somewhat lower amount than what had usually been given out during the last few years. The market interest rate that would have been charged by commercial banks for giving loans for this type of investment was 15%. According to generally accepted accounting practices, the economic life span of the equipment was estimated as seven years.

In a fisheries subsidies study in 2000, the **government cost** of the investment grant scheme was estimated as the total amount of the grants disbursed plus administrative costs (part of the Aquaculture Unit's budget), i.e.: 700 000 + 70 000 = US$ 770 000.

The **value to the industry** in 2000 was calculated as 1/7 of the total amount of the grants received in 2000 (cost allocated over seven years) plus 1/7 of all grants disbursed in the previous six years plus a 15% financial cost:
14 000 000 (total amount of grants 1994-2000) divided by 7 + 15% interest = US$ 2 300 000.

---

and the annual value to the industry should be calculated accordingly. The length of the depreciation period should be based on the expected economic life span of the type of asset in question. We may also want to consider the effect of inflation and the change of monetary values over time and calculate the present value of the depreciation cost for the year of our study.

6.2.3 Vessel decommissioning programmes

Programmes involving financial transfers for reducing fishing capacity are used in many countries. These schemes involve financial compensation for scrapping or exporting fishing vessels to third countries. The effects on the profitability of the industry of such programmes are complex and depend on how exactly the scheme is designed and implemented. If the scheme is used to facilitate the exit – for example the retirement – of individual fishers from the industry, the benefits will accrue to the remaining operators through the sharing of existing resources between a smaller number of actors and thus improve their productivity and profitability – assuming that the decommissioning grant is not at all used for reinvestment in the sector and that there are no new entrants into the fishery, replacing those that left. In the longer term and if the decommissioning scheme has entailed a sustainable decrease in real fishing effort, the effect may also be felt through better catches thanks to an improved resource base. This scenario is of course assuming that overcapacity and overfishing were problems in the first place.

If the decommissioning grant is instead reinvested in the sector, the subsidy would mean a more direct capital injection into the industry that can be used either for covering operating expenses or for other investments and could be considered a **Category 1** subsidy. The effect on productivity and catch volumes will depend on the impact of these expenditures on the total fishing effort and the state of the resources. If the decommissioned vessel is not scrapped but transferred into another fishery, its effect on this other fishery also has to be considered in order to assess the total effect of the scheme on the fishing industry as a whole.

Accordingly, a vessel buyback or scrapping scheme can have a value to the fishing industry in several ways depending on the characteristics of the particular program. There are values created to the industry in the form of the increased resource base left for the remaining fishing vessels to exploit in addition to the compensation payments paid for the scrapped or exported boats. The direct financial transfers made in connection with buy-back programmes are classified as **Category 1** subsidies. If the price paid for the vessel by the government scheme is higher than the market price that could have been obtained had the vessel been put on the market, this surplus constitutes the value of the subsidy to the industry. The cost to the government would be the actual payments plus any related administrative cost. The more implicit resource related effects are better reviewed in the context of **Categories 3 and 4** subsidies. These effects can be immediate or in the longer term and are related to the value of free access to resources. There can be either an explicit increase in quotas for these operators or an implicit possibility to catch more fish thanks to less competition.

Licence, permit and quota buyouts are similar schemes likely to have comparable effects as the decommissioning programmes, depending on the particular conditions and circumstances. In the processing sector, equivalent schemes exist for factory rationalization, i.e. incentives for reducing capacity. Retraining programmes – for fishers or other employees of the sector – with a view to facilitate their redeployment in other industries, i.e. outside the fisheries sector, are measures that also aim at reducing the capacity of the sector.

6.2.4 Equity infusions

Depending on circumstances, we may want to consider government provision of equity capital as a **Category 1** subsidy. If the economic system of the country features a high degree of public intervention in the productive sector in general, partly or fully state-owned enterprises also in the fisheries sector – e.g. hatcheries, ship wharves or fishing companies – would be normal and state capital equity infusions should probably not be included in our fisheries subsidies study. Also, if the state capital investment is consistent with usual investment practices and is made on commercial terms, there is likely to be no cost to the government nor any value to the industry – comparing the terms on which the state investment is made with the conditions of the capital market – and hence the event cannot be defined as a subsidy. However, the issue of state capital equity and state-owned enterprises is complex and further definitions are proposed in Box 6.

Box 6: Defining state capital equity subsidies

To decide if a government equity infusion should be considered a subsidy, we usually look at whether the investment has been made on commercial terms or not. However, the situation can sometimes be confusing and to facilitate the assessment of state capital equity subsidies, the below procedure is suggested. The criteria we are looking at include whether the receiver is a company or not, whether the investment is in the form of equity and if it is made on commercial terms.

Step 1. Define whether the receiver of the investment is a company or not (as opposed to a government institution or department):
It is a company if:
- *It carries out commercial activities*
- *It has a legal form that could also be (entirely) private*
- *It is a tax payer*

Step 2. Define whether the investment is in the form of equity or not (as opposed to a loan or a grant):
It is equity if:
- *It appears in the balance sheet as equity of the company in question*
- *It appears in the balance sheet of the public accounts as a non-depreciable asset (e.g. as a share holding)*

Step 3. Define whether the investment is commercial or not (as opposed to for non profit reasons):
It is commercial if:
- *Return on the investment is required (and dividends – or similar – have been received by the state during the last 5-year period)*
- *The investment itself has been made on commercial terms and is consistent with private sector investment practices*

With regard to the assessment of the subsidy, it is suggested that:

1. If the investment has been invested in a company as equity for commercial purposes:
The **cost to the government** is nil except for administrative costs.
The **value to the industry** equals the financial opportunity cost, i.e., how much would it have cost to borrow the same capital? The annual value of the subsidy is calculated as the estimated interest cost at market rates for a loan equalling the state equity of the company.

2. If the state capital has been invested in a company as equity but not for commercial purposes:
The **cost to the government** is the actual amount of the capital invested (to be reported in the year of disbursement in the same way as a grant subsidy) plus administrative costs.
The **value to the industry** is the financial opportunity cost.

3. If the state capital has been invested in a company but not as equity and not for commercial purposes:
The **cost to the government** is the actual amount of the capital invested plus administrative costs.
The **value to the industry** is the actual amount of capital invested as well as the financial opportunity cost (i.e. the same as a grant).

4. If the state capital has not been invested in a company and not as equity and not for commercial purposes it is not a state capital equity subsidy measure (but may constitute another type of subsidy).

6.2.5   Income support and unemployment insurance programmes

There is a variety of income support schemes and unemployment insurance schemes for fishers. Some of these are part of general social insurance schemes while others are specifically designed for fishers. The schemes can be co-financed by contributions from the fishing industry or be publicly funded. Some examples are:

> ➢ Bad weather unemployment compensation scheme
> ➢ Off-season unemployment insurance
> ➢ Vacation support payments
> ➢ Minimum basic wage
> ➢ Government funded health programmes

Generally, these schemes belong to **Category 1** and their value to the industry correspond to the difference between the actual income fisheries employees receive with the schemes as compared to how much they would have received without them. Industry contributions – or contributions directly by the employees, other than ordinary income tax or other obligatory fees not specific to the fisheries sector – should be deducted to arrive at the net value of the subsidy to the industry.

Box 7: Income guarantee scheme - An example

In our invented country Seidisbus, the fishers in the semi-industrial and industrial fisheries are organized in a Fishers' Association that administers various matters on behalf of the fishers and represents them in different contexts. The Association collects fees from its members to pay for its running costs but there is also a number of support schemes for which the Association receives funding from the government. For example, there is an Income guarantee scheme that compensates fishers for loss of income during periods when fishing fails that is financed at 90% by the state. The scheme guarantees a monthly income of US$ 500. During the year 2000, a total of US$ 500 000 was paid out under the scheme.

In the 2000 fisheries subsidies study, the **cost of the scheme for the government** was calculated as 90% of 500 000 = US$ 450 000. No overhead or administrative cost is considered because the Association manages the scheme and the administrative cost of the Ministry of Fisheries for disbursing the funds is minimal.

The **value of the scheme to the industry** is considered equivalent to the US$ 450 000 received from the government and disbursed to the fishers.

6.2.6 Price support

Market price support can take several forms and is defined by OECD as occurring when the domestic price of a product is higher than the world price as a result of government policy (OECD 2000). Price regulation systems such as those in place in, for example, the EU and Norway are **Category 1** subsidies. Through these systems, compensation is given to fishers whose fish does not reach an established norm price. Price compensation systems can be financed, at least partly, by the industry itself through levies on landed fish. The subsidy's value to the industry is the actual compensation paid out assuming it equals the difference between the amount the fisher would have received had there not been a price support scheme and the total amount the fisher has received with the scheme. If the program is co-financed by the industry, the industry contributions should be deducted in order to arrive at the net value of the program. If a government body

---

Box 8: Positive and negative subsidies

Care should be taken to understand situations where positive and negative subsidies are combined. When there are industry contributions to support schemes and subsidy programmes, these should be deducted from the gross public cost and from the calculated industry value to arrive at the net public cost and the net value to the industry. However, it is usually good practice to show both the gross and net values in a fisheries subsidies report.

administers the program, the cost of the scheme to the government should include an estimated administration cost in addition to the total compensation payments.

A price support could of course also take other forms and could concern, for example, inputs to the fishing or aquaculture subsectors, e.g. drugs needed for breeding of fish or support for the production of a certain gear. Price support also exists with regard to transport costs, often with the objective to reduce disadvantages in remote areas and forming part of regional development programmes.

6.2.7   Export incentives and other market interventions

The marketing side of the industry – both regarding domestic sales and exports – can be supported in many other ways other than by direct market price support, e.g.:

  ➢ Organization of national markets and provision of related infrastructure
  ➢ Regulations in national markets (e.g. sanitary and health standards, obligation of sales through auction, ban on using certain species for reduction, etc.)
  ➢ Aid granted to domestic marketing
  ➢ Sales tax exemptions
  ➢ Promotion of fish consumption
  ➢ Direct export incentives
  ➢ Export and marketing assistance, e.g. measures assisting foreign trade such as international trade fares or provision of information on international markets
  ➢ Market research
  ➢ Free trade zones

These measures are generally **Category 1 or 2** subsidies and should be categorized according to whether the particular support program involves a direct financial transfer to the industry (Category 1), or not (Category 2). Some activities may be classified as **Category 3 or 4** subsidies, e.g. certain types of market regulations.

When assessing the value of the subsidies, there are many different aspects to consider. For example, the organization of national markets probably involves administration – i.e. personnel and overhead costs – but it may also include the physical infrastructure in the form of fish markets. Larger infrastructure projects clearly targeting the fisheries sector

---

Box 9: Depreciation costs

In some countries, there are guidelines and standards for how annual depreciation plans should be calculated based on estimated economic life spans for various types of investments and capital expenditures are allocated over time with annual depreciation costs reported in the public accounts. When available and found to reasonably reflect the likely economic depreciation, we should use these accounting standards for estimating the government cost of fisheries subsidies containing fixed assets. However, in other countries, the government accounts are based on a cash accounting principle and do not include costs for depreciation; the investment is accounted for in its totality at the time of expenditure. In other cases, the depreciation cost reported in the accounts is not based on the expected economic life span but is an accounting or fiscal depreciation cost. This may make it difficult to estimate the annual depreciation cost for our fisheries subsidies study. If the investment is small, the capital cost may have to be disregarded and when assessing, for example, a small landing site, only the operational expenditures as reported in the government accounts would be included. However, when assessing more substantial infrastructure subsidies, e.g. a port facility, it would be necessary to find out – or at least estimate – the relevant investment cost or the real government cost of the measure would be – maybe significantly – undervalued. It is also important to note that earlier subsidy schemes may still have a value to the industry even if no disbursements were made in the year for which the study is being carried out. In the same way, important investments made in the year under study could be allocated over several future years. For more important investments with a long life-span we may also want to consider the effect of time and estimate the present value of the depreciation cost.

such as fish markets – as well as fishing harbours discussed below – are generally examples of public investment subsidies to the fisheries sector. The annual cost to the government of this type of subsidy should be estimated as the depreciation cost per year in addition to maintenance and other operational costs for running the facility. Regarding the value to the industry, it should preferably be estimated as the prevailing market price for using the same type of facility that is being provided. In many cases, however, there is no market price alternative and we may have to use the government cost as a proxy when estimating the industry value.

Government activities that indirectly support the marketing side of the fisheries sector are difficult to assess. For example, the promotion of fish consumption could be part of a broad government information campaign for healthier food habits and the fisheries sector would then only be one among other food sectors being affected. An estimate of the value of the campaign to the fisheries industry should then be based on only a part of the overall cost for the campaign. This cost accruing to the fisheries sector could be calculated according to a distribution index based on the total value added created by the different subsectors, i.e. generally reported as the different subsectors' contribution to GDP[7]. Other distribution index that could be used, depending on circumstances, include, for example, the number of employees or the total sales value (turnover) in the various subsectors.

6.2.8 Import quotas, tariffs and other border measures

Border measures that do not involve a financial transfer to – or from – the industry can be classified as **Category 2** subsidies. These include regulatory frameworks such as import quotas and other non-tariff measures, import tariffs as well as landing bans for foreign fishing vessels and can represent important advantages for the domestic industry. The measures represent in practice transfers from consumers to fishers arising from government policy (Flaaten and Wallis 2000). Tariff escalation regimes are border measures that benefit in particular the processing industry by allowing raw fish to be imported at lower tariffs than processed products. For importers and traders selling imported products, tariffs may instead constitute a negative Category 1 subsidy if the import duties on fishery products are higher than on other imported goods, in particular foodstuff.

Border measures are often difficult to assess with regard to their value to the industry. If there are international prices available for the products in question, these prices could be used in a comparison with domestic prices to assess how the measure has influenced the national market and price structure. If there is a difference between local and international prices that cannot be explained by other influences, this difference could be used for drawing conclusions with regard to the border measure's impact on, for example, revenues to the local processing industry. However, one of the reasons why PSEs (see Box 4) are difficult to calculate for the fisheries sector is that the wide variety of processed fishery products, many of which are market specific, do not have internationally traded equivalents. It would also be difficult to assume that there are no other influences and the calculation may require more statistical data than are available.

---

[7] It should be noted that a measure of the whole fisheries sector's - as defined by the Guide - contribution to GDP may include contributions from subsectors not directly reported as fisheries, e.g. the food processing industry and the mechanical industry.

Box 10: Calculating the value of a fish export promotion subsidy – An example

In Seidisbus, exports of fish and fishery products represented 25% of the total export value in the year 2000. Under the Ministry of Trade, there is an Export Council working on promoting exports, both of fish and of other products. In addition to providing various information and contact services, the Council organizes a trade fair every two years. The fair is funded by the Council and by contributions from the participating export companies. In the trade fair 2000/2001 – held in February 2000 – 30 of the 100 exhibitors were from the fisheries industry.

In the fisheries subsidies survey, the activities of the Export Council were found to be a subsidy to the export sector as industry focusing on the domestic market did not receive the same support. The total budget of the Export Council in 2000 amounted to US$ 300 000 covering costs for personnel, rent of offices, depreciation on vehicles and office equipment, and other operational expenses. The cost for organising the trade fair in 2000 was US$ 50 000 whereof US$ 40 000 were extra-budgetary funds provided by the Council (i.e. excluding costs for ordinary staff of the Council and general overhead costs) and US$ 10 000 fees collected from the participating companies.

The **cost to the government** of the "Export Council subsidy" in 2000 was calculated as follows:
Cost of general activities in 2000:
25% (fisheries' share of exports) of 300 000 (total budget) = 75 000.
Cost of trade fair 2000/2001:
30% (fisheries industry participants) of 40 000 (Council contribution) = 12 000
TOTAL: 75 000 + 12 000 = <u>US$ 87 000</u>.

Considering that the trade fair is a biannual event, we could consider allocating the cost over two years. However, in this particular case, it was decided against considering the relatively small amount and that the benefits of the fair were likely to occur already in 2000.

With regard to the **value of the subsidy to the industry**, no comparable market prices were found and it was decided that the cost to the government should be used as a proxy for the industry value, <u>US$ 87 000</u>. Over the next year, the Export Council plans to do a questionnaire survey among the exporting companies to evaluate the likely impact of its activities on export sales and net income. In a future fisheries subsidies study, the results of this survey could be used for assessing the value of the subsidy to the industry.

There are also border measures such as restrictions with regard to foreign direct investment, both in the processing and harvesting subsectors. Restricting competition through stopping the free movement of harvesting services constitute implicit assistance to those fishers who are allowed to fish. Such measures include *inter alia*:

> Ownership restrictions
> Allocation of catch quotas only to national fishers
> Nationality and residence requirements for company officials/managers and crew

The effect of these types of subsidies to the domestic industry is translated into less competition and therefore potentially larger market shares. When attempting to assess the value of these larger market shares, we are likely to encounter similar difficulties as when assessing other border measures.

To use the government cost of border measures as a proxy may not either appear as a satisfactory option. The cost is likely to be a fairly low administrative cost, badly reflecting the true value of the measure. The government cost may also be difficult to estimate correctly because it is likely to require various overhead calculations. Another option is to use a pro

---

Box 11: Administrative costs and overhead

Many of the costs related to fisheries subsidies are indirect costs, i.e. overhead costs of authorities and administrations implementing a support scheme or a regulation. By definition, these costs are generally not directly related to a specific activity and need to be calculated using some sort of distribution index as discussed on page 47 above. In this context, we also need to think about how many different stages or levels of overhead we want to include when assessing a particular subsidy. Let us assume that we are calculating the government cost for an investment grant scheme in the aquaculture sector (see also Box 5). The scheme is administered by the Aquaculture Unit of the Department of Fisheries of the Ministry of Agriculture and Natural Resources but financed from a special fund for sustainable rural development within the Ministry of Finance. The cost of the program is first of all the grants themselves, which represent the direct cost. But what administrative and overhead costs should we include? If the administration of the scheme makes claim to a fair amount of time and resources of the Aquaculture Unit, part of its budget – calculated according to a suitable distribution index – should be assigned to the grant scheme. Then we could argue that also part of the budgets of the Department of Fisheries and the Ministry of Agriculture and Natural Resources should be linked to the grant scheme as the Aquaculture Unit draws on their resources. Some administrative costs of the Ministry of Finance could also be related to the channelling of funds. However, these calculations easily become quite complex and if we think that the administrative cost is negligible, we could probably ignore it. Accordingly, in the above example, only calculating the overhead and administrative cost of the Aquaculture Unit may satisfy us.

---

rata standard value, valuing the support measure at, for example, a fixed percentage rate of turnover. The particular situation and the framework of the actual study will have to decide which approach that we should choose.

Box 12: Fuel tax rebate - An example

Fishing vessels in Seidisbus that are registered with the Fisheries Department benefit from a fuel tax rebate. The amount of rebate depends on the type of fuel used; gasoline, diesel and oil mixtures, and there are maximum quotas per annum per fishing vessel based on the horsepower of its engines. The rebate is refunded by the Department of Fisheries when the eligible fisher presents a claim.

The **cost to the government** is calculated as the foregone revenue plus administrative costs, i.e.:
US$ 0.07 (average rebate per litre of fuel) multiplied by 10 000 000 litres (amount of fuel for which claims have been made) = 700 000.
Administrative cost: 10 000.
TOTAL: US$ 710 000

The **value to the industry** is assumed to be the same as the actual rebate, i.e. US$ 700 000.

Even though tax rebates are generally classified as **Category 2** subsidies, we could consider classifying this particular program as a **Category 1** subsidy as it in practice involves a direct financial transfer (through the procedure of reimbursements according to claims).

6.2.9 Fuel tax exemptions

One quite common fisheries subsidy is to provide fuel at a lower tax rate to fishing vessels. If the fisheries industry has access to fuel at a lower cost than other industries, the scheme could constitute a **Category 2** subsidy.

The value of tax exemptions is calculated as the difference between the "normal" tax rate, i.e. the rate applied to other sectors of the economy, and the lower rate granted the fishing industry. For the government, the cost would be represented by the foregone tax revenues and administrative costs related to the scheme.

Other reductions in public fees and taxes implying that inputs, supplies and services are provided below market price should be valued at the difference between the price actually paid by the industry and the market price or price generally charged to other sectors.

6.2.10 Investment tax credits and deferred tax programmes

Benefits gained through investment tax credits should be assessed by comparing the subsidized scheme with the normal tax regulations applicable to other industries. However, because this type of tax credit often means a redistribution of costs over a period of years by allowing accelerated depreciation of fixed assets, i.e. faster than the real economic life span, or by allowing investments to be made out of non-taxed profits on certain conditions, the actual value of the scheme to the industry in a specific year is usually difficult to calculate. One benefit is the extra capital made available for additional investments and this could be valued at the cost of commercial interest rates. Other benefits include the easing of fluctuations in income over a period of years that would constitute a subsidy equaling, for example, an income loss or unemployment insurance or the financial cost of borrowing working capital.

Deferred tax programmes are similar to the investment tax credits and a similar approach for evaluating their benefits to the industry should be applied. With regard to government costs, it is the foregone revenue that should be estimated.

6.2.11 Favourable loans and loan guarantees

When the fisheries industry is offered loans on favourable conditions through government institutions, these are often classified as **Category 2** subsidies. A favourable loan may be a loan at a subsidised interest rate or on other favourable terms such as an extended amortization period. When there is a subsidised interest rate – or a favourable interest rate is obtained with the help of a loan guarantee – the value to the industry could be estimated by comparing the subsidised interest rate with prevailing market rates. When the subsidy consists of other favourable terms, the assessment of the value to the industry becomes more difficult and will have to be reviewed on a case by case basis. With regard to the government cost, it would be appropriate to consider costs related to payments for defaulted loans. If there are no such costs, the cost to the government is usually limited to the administrative cost of operating the schemes.

6.2.12 Special insurance schemes for vessels and equipment

Insurance schemes run or underwritten by governments are often classified as **Category 2** subsidies when they offer the fisheries industry terms and conditions that are more favourable than those on the commercial insurance market. The industry value of these schemes could be estimated as the difference between the subsidised premium cost to the industry and the corresponding market price for an equivalent insurance. If there is no market price available for the particular type of insurance, an approximation could be made taking the perceived risk into account. The government cost would be calculated as the amount of claims paid out and the administrative cost involved in the managing the scheme less insurance premiums paid by the industry. Also the value of the subsidy to the industry could be based on these actual government costs, in particular if the amount of claims is significant and there are no applicable market prices.

6.2.13 Training and extension

Various types of specialized training and extension services are sometimes available for the fisheries industry, fully or partly funded by the government. There may be training courses on fish handling, safety at sea or in seamanship. Many governments provide extension services with a view to facilitate, for example, the introduction of technologies in the processing sector or to promote the use of better seeds in rural aquaculture. Improved skills usually mean improved production with increased income as a result. It such an effect can be deduced directly from the subsidised training or extension scheme and we can measure the effect, this could also be used as for the assessment of the value to the industry.

To assess the cost to the government of these services may be fairly easy, in particular if they are provided through separate administrative entities, such as a training institute or an extension unit of a fisheries division. With regard to the value of training courses to the industry, market prices may be available for similar tuition in other subject fields.

6.2.14 Inspection and certification services

Strict quality requirements in the world's main fish importing countries have put pressure on quality assurance for export products. European importers (the EC; now the EU) issued the first regulations with regard to the control of fishery products in 1991 and have since expanded the system to the so called "own health checks" which extends the application of hygiene and quality controls to the whole production chain. Also other countries have introduced similar regulations (FAO 2000b). If the required inspection and certification services are provided to the exporting industry free of charge or at a price lower than the related operational costs, we may want to classify the services as a **Category 2** subsidy.

The production standards stipulated by the importers also generally require investments in equipment and infrastructure. If these investments are paid for by the industry, the regulation may initially have a negative impact on the industry's profits and only pay off in the medium or longer-term. It could hence be classified as a **Category 3** subsidy. The assessment of the value to the industry of the regulation and the services provided should preferably be based on an estimate of the value of the increased exports to the markets requiring the certification and the costs of fulfilling the conditions.

6.2.15 Fishing port facilities and other infrastructure

Governments generally provide infrastructure such as roads, dams, bridges and public buildings and this is commonly considered to be the responsibility of the government: it is acceptable to finance basic infrastructure, beneficial to citizens in general, by tax payments through the public budget (at least partly; there are also many examples of users contributing directly to the costs of some of the more general facilities through, for example, road taxes).

However, infrastructure that is specific to a group of citizens or a particular sector of the economy and for which the costs – investment costs and operating costs – are not recovered from these groups of users could be considered a subsidy. The line between general and specific infrastructure is sometimes difficult to draw and we have to refer to the macro-economic framework and common practices in our country of study to decide what infrastructure should be considered fisheries subsidies. Commonly, one of the more clear examples of a subsidy of this type is the provision of fishing port facilities. Harbour dues are often collected but unless these cover the entire cost of building, maintaining and running the port, the provision of the facility could be considered a **Category 2** subsidy. Other examples of fisheries specific infrastructure are fish markets that were mentioned under Export incentives and other market interventions above. The principles for assessing the costs and values of harbour or port facilities are the same as those applicable to fish markets.

Box 13: Assessing infrastructure subsidies - An example

There are ten different landing sites and small ports along the coast in Seidisbus, operated by the local city municipalities. The facilities at the landing sites vary but generally include gas pumps, jetties, toilets, storage lockers, fish wash stands and engine repair rooms. At each site, there is a caretaker – employee of the municipality – who is responsible for the management of the site. The sites are mainly used free of charge by some 800 artisanal fishing boats.

In a fisheries subsidies study, the provision of these landing site facilities free of charge was considered a subsidy to the fisheries industry as this type of services does not generally exist in other economic sectors. However, the assessment of the subsidy caused problems because the capital cost of the investments were not accounted for on an annual basis in the accounts of the municipalities; as the government practices a cash basis accounting principle, the capital expenditure is accounted for in the year of payment and no depreciation costs are allocated over time. Moreover, some of the facilities were very old and it seemed difficult to establish when they had been constructed. Hence, to estimate the annual depreciation cost to be included in the calculation of the government cost of the subsidy various assumptions and approximations had to be used. Some of the investment expenditures could be found in the accounts of previous years. By consulting other departments of the government involved in public infrastructure projects, estimates of the values of other items were obtained as well as their likely economic life spans.

Accordingly, the total **cost to the government** of the ten landing sites were calculated as follows:

Operational costs: 100 000 (ten caretakers) + 50 000 (maintenance and repairs) + 20 000 (miscellaneous) = 170 000.
Depreciation on infrastructure type 1: 200 000 (investment) divided by 10 years = 20 000.
Depreciation on infrastructure type 2: 300 000 (investment) divided by 20 years = 15 000.
Overhead cost city municipalities: 5 000.
TOTAL: US$ 215 000

With regard to the value of the subsidy to the industry, the annual cost per boat, i.e., approximately US$ 270 according to the above calculation, was compared with the prices charged by two private boat clubs offering mooring and other facilities to leisure boats: US$ 500 per boat per year. Taking the differences in facilities and services into account, it was concluded that a reasonable market price for the landing site facilities offered by the City Municipalities would be around US$ 350. Hence, the **total value to the industry** was calculated as:
350 multiplied by 800 (number of boats) = TOTAL: US$ 280 000

6.2.16 Payments to foreign governments to secure access to fishing grounds

In some countries, domestic fishers are granted the privilege of free access while foreigners pay some sort of access fee. When a government pays these access fees in foreign countries for its own fishing fleet, these could be considered as **Category 2** subsidies with a value to the industry equivalent to the actual annual cost of the fishing right.

6.2.17 Government research and development (R & D)

Governments often fund research institutes and activities. Certain R & D, leading to efficiency improvements, is likely to be carried out also by the industry and government funded research is then a direct support to these activities. Other research may be more related to fisheries management and resource protection and could, for example, provide management information or lead to the development of gear that is then imposed on the industry through gear regulations. R & D activities are probably best classified as **Category 2** subsidies. With regard to the assessment of their costs and values, we are likely to encounter similar difficulties as we are for fisheries management and the industry value may have to be

---

Box 14: Delineating subsidies, assessing cross-departmental activities and public accounting practices

Sometimes it may be more convenient to define a subsidy according to the organization that delivers the subsidy rather than splitting it up in actual activities or support measures. For example, a research or training institute may be defined as a subsidy in its entirety, rather than separating out the research activities or training courses. At other times, the subsidy is better reported as an activity with several providers, e.g. as fisheries management including inputs from a management division of the Ministry of Fisheries, research activities from an institute and surveillance carried out by the Coast Guard. The approach we chose would depend on the level of detail we are aiming at for our fisheries subsidies study but there are also practical considerations. The way we delineate our subsidies may have to be influenced by the structure of the public accounts and the organization of the public bodies providing the subsidies. It may be very difficult to calculate the cost (revenue) to the government of a certain situation or measure if the subsidy is defined in a way that cuts across several departments or accounting categories.

On the same note, we may want to carry out the study for a financial year, and not on a calendar year basis, unless these coincide. When there is a choice of using the public budget allocation for the year of the study or the actual expenditures incurred, we can use either of the two approaches as long as the difference between the two is not important and a consistent method is used. Likewise, we may have to decide whether we should use figures for approved expenditures or actually disbursed funds. For example, a number of investment grant applications may have been approved in December – at the end of the financial year that we are studying – but the payments are only going to be effected in January. Again, we can use either the accounting or cash basis for our assessment, depending on how the public accounts are organized.

In the same context, care should be taken when assessing and reporting on subsidies involving government agencies receiving their funding from other subsidy providing government agencies; we have to avoid double accounting. For example, funds for training provided by the Ministry of Fisheries to a training institute should not be reported both as part of the Ministry's subsidy costs and those of the institute.

---

assumed to be the same as the government cost (see below).

6.2.18 Fisheries management and environmental protection programmes

Fisheries management is one of the more complex areas with regard to subsidies. The activities included in fisheries management can be divided into three main areas covering a variety of measures and programmes, i.e.:
- Research for the creation of information for management decisions
- Design and implementation of fisheries management systems
- Enforcement of fisheries management rules
  (Wallis and Flaaten 2000).

Governments commonly spend a lot of money managing their fisheries. According to an OECD study, 4.9 billion US$ or 77% of all government financial transfers to the fishing industry in 24 OECD countries in 1997 were spent on "general services", a definition covering the three activities included in fisheries management (OECD 2000)[8]. In a handful of countries, fisheries management costs are recovered from the industry but in most countries the service is paid for by the fisheries administration and financed via the public budget. In accordance with the Guide's definition of fisheries subsidies, the national context will decide whether we should consider fisheries management a service that should be provided free of charge – and hence not defined as a subsidy – or whether it is subsidy as the costs are not recovered.

Fishers are likely to benefit from fisheries management by the increased long-term sustainable output from the fishery that a successful management system implies. In the short and medium term, there are also benefits from reduced competition – when access is restricted – and from, for example, regulations allowing fish to grow to larger sizes and hence increasing the return per unit of output (Wallis and Flaaten 2000). However, in the short term, fishers may also experience management negatively if it restricts catch volumes.

Accordingly, different fisheries management activities can be classified into different subsidy categories. Most would probably be **Category 2** but others – having both a short-term effect (often as a negative subsidy) and a long-term effect (as a positive subsidy) – are probably better classified as **Category 3** subsidies. The lack of appropriate regulations is generally a **Category 4** subsidy, benefiting the industry in the short-term but with possible negative consequences in the longer term. It is, however, recognized that it may be difficult to separate out different parts of the management system in this way in practice.

Examples of support programmes and measures that could be identified as subsidies – of different categories – related to fisheries management and to environmental protection include:
> Compensation schemes (for closed season, for damaged fish stocks, gear conflict, temporary withdrawal): **Category 1**
> Support to community based management and regional development bodies: **Category 2**
> International cooperation including payment for membership in international and regional organizations active in the field of fisheries management: **Category 2**
> Fisheries enhancement and stocking programmes: **Category 2**
> Support to consultative groups and mechanisms: **Category 2**
> Gear regulations (e.g. Turtle Excluder Devices): **Category 3**
> Chemical and drugs regulations for aquaculture: **Category 3**

The basic characteristics of fisheries management services – i.e. being activities that the industry would not be in a position to organize and implement itself[9] – make it difficult to estimate a market price equivalent for this type of support. Hence, we are likely to have to resign ourselves to using government costs as a proxy for the value of fisheries management

---

[8] It should be noted that the OECD definition of government financial transfers is narrower than the Guide's definition of subsidies and does not, for example, include market price support measures. The study also only covers the capture fisheries subsector. Still, the proportion of funds spent on fisheries management is considerable.

[9] There are several reasons why it cannot be expected from the industry to undertake fisheries management, i.e.: the industry lacks the legal authority required to make fisheries regulations successful; the industry is generally not authorized to ban new entrants or impose restrictions on operators; the public good character of the fishery resources invites to free riding behaviour of individual members of the industry (Hannesson 2000).

Box 15: Membership fees to international organizations - An example

Our invented country Seidisbus is a member of several international and regional organizations of which two are of relevance to the fisheries sector: FAO and a regional fisheries committee for the management of small pelagic stocks. FAO has an office in the country and is also currently implementing a project on marine fisheries management in cooperation with the government. The committee has recently been established and only one meeting has been held so far.

In the fisheries subsidies study, it has been agreed that these memberships and their related activities should be considered subsidies to the fisheries industry. The subsidies are related to fisheries management but are reported on separately in the subsidies studies, partly because of their nature and partly because their costs to the government are relatively easy to locate in the public accounts.

The annual **government costs** for the different activities are calculated as follows:
Membership FAO: 15% (fisheries share of agricultural GDP) of 500 000 (Seidisbus' annual contribution to FAO) = 75 000.
Hosting of FAO office: 15% (fisheries share of agricultural GDP) of 45 000 (annual operating expenses for the office assumed by the government) = 6 750.
Management project: 20 000 (annual counterpart contribution).
TOTAL FAO: US$ 101 750

Membership regional fisheries committee: 50 000 (annual fee).
Travel costs to meeting for government representative: 5 000 (estimate based on air ticket prices and government DSA rates).
Administrative cost: 10% (estimate) of 300 000 (budget of the International Cooperation Unit of Department of Fisheries) = 30 000.
TOTAL Regional fisheries committee: US$ 85 000

Given that no market price or estimate of the impact on industry income is available, the **value to the industry** of the activities is estimated to be the same as the government cost with the addition of the FAO contribution to the fisheries management project, i.e.:
TOTAL FAO 101 750 + 100 000 (annual project budget) = US$ 201 750
TOTAL Regional fisheries committee: US$ 85 000

---

to the industry. This is maybe in many ways not a satisfactory way of assessing the presumed short- and long-term effects of catch restrictions or resource sustainability but it is probably the only practical measurement available. The net value of the subsidy equals the total government expense minus any cost recovery form the fishing industry. Cost recovery could take the form of user fees or similar levies. It is important to assess the net value of fisheries management in relation to the costs and value of resource access – discussed below – because the two are closely related.

6.2.19 Free or below market price access to fishing grounds

To allow fishers to use resources free of charge or at a cost lower than the actual value of the resources could constitute an important cost-reducing subsidy to the industry that would be classified in **Category 4**. This is true both in cases of open access as well as in management systems by which permanent or temporary quotas are transferred to the industry, e.g. ITQs. The argument is based on the reasoning that all resources that are scarce have a value to society and by making them available – either indirectly or by granting formal user rights – to the direct users without charging for the use must be considered a subsidy[10].

---

[10] There are of course different views on this and the arguments are related to the concept of property rights (Schrank and Keithly Jr. 1999).

Box 16: Gear regulations - An example

In 2000, the use of Turtle Excluder Devices (TEDs) became obligatory on all vessels fishing in the waters of Seidisbus. Many boats fishing for export markets were already using TEDs but the usage was now introduced on a wider scale. The government launched an information campaign and also provided advice free of charge to fishers who were, however, expected to pay for the TEDs themselves. Extra resources were also allocated as from the 2000 budget to the Department of Fisheries for inspection and control of the proper use of the device.

The **cost to the government** in 2000 for introducing the measure was:
10 000 (information campaign and advice services) + 20 000 (inspection and control) + 5 000 (overhead Department of Fisheries) = US$ 35 000
It could be argued that the initial costs should have been allocated (depreciated) over a period of several years as they represent a onetime lump-sum, i.e. as an investment at the beginning of the scheme. However, given that the amount was rather small, this was not considered necessary (see also Box 9).

Regarding the **value of the regulation to the industry**, the initial impact in 2000 is negative. The marine capture subfisheries sector invested in a total of 100 new TEDs at a cost of US$ 500 each. No immediate economic benefits were perceived even though increased sales in export markets could be expected in the medium-term thanks to improved credibility. Hence, the value reported in the fisheries subsidies report in 2000 was - US$ 50 000 (negative).

This concept of free access to resources should not be confused with the issues of fisheries management, discussed above, even though the two are very closely related. The provision of fisheries management could be a subsidy per se and, in addition, there can be a "resource subsidy". In a hypothetical situation without any fisheries management and hence free access to resources, the latter is a subsidy. Where there is a fisheries management system in place and fishing quotas are distributed to the fishers free of charge or at a nominal fee, the industry could be said to benefit from both a "fisheries management subsidy" and a "resource subsidy"." At the same time, it should be pointed out the two issues are interrelated – for example, a quota in a well-managed fishery is likely to be worth more than in a badly managed fishery. Moreover, user fees are often used as a management tool.

The support measures categorized in **Categories 3 and 4** are the most difficult ones to assess the value of. Still, with regard to fisheries management, discussed above, there are the actual costs to the government for the existing management system that can be used as a basis for estimating a value to the industry of the services provided. For **Category** 4 subsidies, there is – as we have already noted – no government cost involved and this proxy value is hence not available to us. Moreover, fisheries management and resource access regimes are in practice closely related as access fees are usually seen as a mechanism for recovering management costs. This complicates the matter even more. How do we estimate a value of the resources?

According to economic theory, the value of the resources could be estimated as the opportunity cost to society – or shadow price – of making the fishery resources available to the fishing industry in a certain way. The user fees could be set at a level allowing the government to recover the society's full cost, i.e. including costs associated with the impact of fishing on non-targeted species, with collateral environmental impacts and the more general cost of removing the resource at present rather than in the future (Milazzo 1998). However, it could also be argued that, if environmental and social costs are disregarded, the immediate opportunity cost of, for example, a free distribution of quotas corresponds to the price that the government could obtain if making the fishery resources available in an alternative way, e.g. selling the rights to fish in an open market to the highest bidder. The value to the industry would also correspond to this market price, i.e. the price that the industry would be prepared to pay for the quota. Unfortunately, there are very few examples where there is this type of free and open market for fishing rights that could give guidance to the value of the resources in general. Moreover, the existing management regime influences the resource rent of the fishery, which in turn is likely to have an influence on the price the individual operators would want or be able to pay as user fees. With this circle of causes and effects, the actual value of the resource is difficult to determine.

Many countries charge licence fees to foreign fishers but not for domestic operations, or at a much lower level. This difference could be used as an indication of the value of the fisheries resources, i.e. the subsidy to the domestic industry would correspond to the access fee charged to foreign operators minus the domestic fees, if any. However, it would again be difficult to separate the fisheries management component from the resource access benefit. Empirical work shows that there is a wide variation of currently used rates. According to Milazzo (1998), considering the whole spectrum of user fees, both for foreign and domestic fishers, there is a range from less than one percent to more than one-third of the ex-vessel value of the production. Milazzo uses 5% and 10% of ex-vessel values as two different estimates of the global domestic resource rent subsidies (excluding distant water fishing fleets). These figures include both the "fisheries management subsidy" and the "resource subsidy". Here we are at the moment only interested in the "resource subsidy" and in the absence of real values or any international standards, the Guide proposes to use a similar general value for estimating the value of resource access at 3 to 5 percent of the ex-vessel value of production. This value should be added to the costs of management, discussed above, and factual user fees or fishing rights charged should be deducted to arrive at the total of the two subsidies for fisheries management and access rights.

Box 17: Free access to resources - An example

In Seidisbus, there are no licence fees for national fishers. In fact, no licence or access fee is charged to foreign fishing vessels either as fishing by third country only takes place within in the framework of reciprocal agreements with neighbouring countries.

The total value of the fish landed in Seidisbus is US$ 75 millions (see Box 18). As an estimate of the potential value of the free access to resources, 4% of this landed value is suggested. This gives a total of 3 millions which would then correspond to a cost to the government as foregone revenue. However, if this access fee were to be collected, it would involve an administrative cost to the government. The net **cost to the government** of this Category 4 subsidy would thus be:
3 000 000 - 100 000 (estimated administrative cost) = US$ 2 900 000

The current **value to the industry** of the free access regime is the US$ 3 000 000.
In a longer-time perspective, it could be expected that the industry will suffer from a diminution of resources and related problems common to non-regulated fisheries. However, it was not considered possible to quantify these effects in the current Seidisbus study.

6.2.20 Lack of pollution control

If the government allows the fisheries industry to pollute in a way that other industries cannot, we may have identified another **Category 4** subsidy. It is also another subsidy for which the value to the industry is difficult to assess: how should it be decided what the "normal" pollution control level is? In a full cost recovery situation and analogous to the discussions above on fisheries management and user fees, the fisheries sector should pay for all collateral impacts on the environment as well as negative externalities caused to other sectors. In practice, the Guide suggests that the value for this type of subsidies is assessed by comparing the fisheries sector with other economic sectors with a view to identify areas in which the fishing industry is under less control than other activities. The value of not imposing pollution control would equal the cost of alterations – in equipment and in practices – that would otherwise have been required for ceasing the polluting behaviour.

6.2.21 Lack of implementation of existing regulations

Another type of situation that could be defined as **Category 4** subsidies is the lack of implementation of existing rules and regulations. This could apply to, for example, hygiene and quality control that is not carried out or to fees that should be collected but are not.

6.3 Identifying and assessing subsidies – some remarks with regard to categories 3 and 4 and longer-term impact

In the discussion above on how to assess fisheries subsidies, it transpires that while it is relatively easy to measure subsidies of categories 1 and 2, it is much more difficult to identify and assign reasonable values to subsidies from categories 3 and 4. In our examples, the only values that we have been able to estimate for the categories 3 and 4 are short-term. This is unfortunate as one of the distinguishing characteristics of these two groups of subsides is the long-term perspective they allow us to consider.

The approach in the Guide of focusing on the short-term impact does not imply that the long-term effects are considered unimportant. It is recognized that subsidies have much more far-reaching consequences than the immediate change in profits and the current public budget implication. In the longer-term, subsidies will affect the actual structure of the industry through the impact on the economic performance and on the change in behaviour of the actors of the sector that this triggers. This is also often the reason why governments introduce subsidies: to influence the development of the industry by giving it incentives to do things a certain way, which will benefit society. The long-term impact will hence be felt both at the macro-economic level in the form of costs or benefits to the environment and society as well as at the level of the fisheries industry. This is of course also true for the Category 4 subsidies even though they often exist not because conscious decisions have been made to introduce them but because the industry and society itself – and its economic structure – have changed over time creating new situations.

However, with the knowledge currently available, the longer-term aspects are difficult to assess and the Guide's focus with regard to the quantitative assessment remains on short-term effects. The assessment we make constitutes a "snap-shot" of the current situation. At the same time as this may be felt to be insufficient, it is believed that it is an inevitable and important step towards a better understanding of the impact of fisheries subsidies, also in the longer-term.

These issues should be kept in mind when referring to Figure 9 in which the examples of subsidies presented in chapter 6.2 are summarized.

Figure 9: Summary of the assessment of fisheries subsidies in the country Seidisbus[11]

LIST OF FISHERIES SUBSIDIES IN SEIDISBUS (Year 2000)

Name of subsidy	Category	Cost (revenue) to government (US$)	Value to industry (US$)
Investment grants for storage and transport in aquaculture subsector	1	770 000	2 300 000
Income guarantee for fishers	1	450 000	450 000
Fish export promotion through the Export Council	2	87 000	87 000
Fuel tax rebate for fishing vessels	2	710 000	700 000
Provision of landing sites and ports	2	215 000	280 000
FAO membership and project and counterpart contributions	2	101 750	201 750
Membership in regional fisheries committee	2	85 000	85 000
Introduction of obligatory Turtle Excluder Devices	3	35 000	(50 000)
Free access to fish resources by the national fleet	4	2 900 000	3 000 000
TOTAL		5 353 750	7 053 750

7 COSTS AND EARNINGS ANALYSIS – THE IMPACT ON INDUSTRY PROFITS

7.1 The profit and loss account and the classification matrix

In the chapter above, we discussed various issues regarding the assessment of different types of fisheries subsidies. Approaches and methodologies were suggested for how to calculate their cost – or revenue – to the government and their value to the industry. In this chapter, we will look closer at the latter aspect and see how we can analyse the impact of fisheries subsidies on the profitability of the industry in more detail.

The classification of subsidies into four categories gives us a good idea of their modalities and general characteristics. However, to analyse their impact on the profitability of the fisheries industry, we would like to know more precisely in what way they influence the industry's financial situation. Do revenues increase? Do variable costs decrease? Or are capital costs influenced? It is suggested that we add another dimension to the subsidies classification system that will give us information on which revenues and costs that are being affected. Table 1 shows how the different subsidy examples cited earlier could fit into a classification matrix based on revenue and cost types in addition to the categories from chapter 5.

---

[11] The values reported in this table and in the underlying examples are fictive and do not necessarily bear any resemblance to real costs and values.

Table 1: Fisheries subsidies classification matrix

Reference to profit / loss account	Category 1	Category 2	Category 3	Category 4
REVENUES				
SALES REVENUES (AND OTHER INCOME)	*Price support (+) Direct export incentives (+) Vessel decommissioning programmes (+)*	*Import quotas (+ or -) Export promotion support (+) Direct foreign investment restrictions (+) Inspection and certification for exports (+) Fisheries management (+) International cooperation and negotiations (+)*	*Environmental protection programmes and regulations (+LT) Gear regulations (+LT) Chemical and drugs regulations (+LT)*	*Non implementation of existing regulations (-LT) Free access to fishing grounds (-LT) Lack of pollution controls (-LT)*
OPERATING COSTS				
RUNNING (VARIABLE) COSTS	*Import/export duties (-)*	*Fuel tax exemptions (+) Port and landing site facilities (+) Provision of bait services (+)*	*Chemical and drugs regulations (- ST)*	*Non implementation of existing regulations (+ST)*
LABOUR COSTS	*Income guarantee programmes (+) Disaster relief payments (+)*	*Special income tax deductions (+) Specialized training (+) Extension (+)*		
FIXED COSTS	*Grants for safety equipment (+) Taxes and fees (-)*	*Special insurance schemes for vessels and gear (+) Payments to foreign governments to secure access to fishing grounds (+) Government R & D programmes (+)*	*Environmental protection programmes and regulations (-ST) Gear regulations (-ST)*	*Free access to fishing grounds (+ST) Lack of pollution controls (+ST)*
GROSS CASH FLOW				
CAPITAL AND FINANCIAL INCOME AND EXPENSES				
DEPRECIATION AND INTEREST COSTS	*Investment grants (+) Equity infusions (+)*	*Investment loans on favourable terms (+) Loan guarantees (+) Investment tax credits (+)*		
PROFIT OR LOSS BEFORE TAX				
TAX				
CORPORATE INCOME TAX		*Deferred tax programmes (+)*		

ST=short-term effect; LT=longer-term effect; + = positive effect (revenue-increasing or cost-reducing); - = negative effect (revenue-decreasing or cost-increasing)

As we can see, the structure of a profit and loss account has been used as a model according to which the influence on profitability has been divided into revenue-enhancing (or reducing, if a negative subsidy) and cost-reducing (increasing) subsidies. The latter have been further divided into subsidies influencing mainly running costs, labour expenses, fixed costs or capital and financial expenses in order to give an indication of the type of cost that is affected and whether the subsidy affects profitability immediately or in the medium-term: a subsidy implying a change in capital is likely to have a longer-term impact than a subsidy affecting running costs[12]. The definitions of the different revenue and cost groups are:

- *Revenue enhancing (reducing) subsidies*
  The concept "revenue" refers of course firstly to income from ordinary sales but also includes income from other sales, i.e. the disposal of equipment. The revenues are decided by the production volume – catch volume if a harvesting operation – and the price obtained for the goods and hence all subsidies affecting sales volumes or prices would belong to this group.

- *Cost reducing (increasing) subsidies: general running (variable) costs*
  The running or variable costs subsidies are defined as those affecting the operating costs that vary in the short-term with the rate of output. These include, for example, raw material and fuel costs and would generally, in the harvesting subsector, vary with the number and length of fishing trips. Often labour costs are also reported among the running costs but here subsidies related to labour are dealt with in a separate group.

- *Cost reducing (increasing) subsidies: labour costs*
  All subsidies applicable to labour costs have been classified in a separate group. Labour costs include in particular wages and various social insurance payments as well as all human resource development.

- *Cost reducing (increasing) subsidies: fixed costs*
  Fixed costs are costs that do not vary with output in the short term. These often include overhead costs. Subsidies related to costs that are generally fixed for at least one year but excluding measures affecting basic investments and financial costs are classified in this group.

- *Subsidies influencing capital and financial expenses (depreciation and interest costs)*
  Subsidies having an impact on the profitability of the fisheries sector by affecting the costs of investment are classified in this group.

- *Taxes*
  Subsidies related to company income tax are reported in the last of the profit and loss account subsidy groups.

As we noted with regard to the classification of subsidies in the four main categories, there may be subsidies that appear to suit equally well into several different groups and sometimes it can be difficult to decide whether a certain measure affects, for example, variable or fixed costs – or both. We should try to determine the most direct, immediate and significant impact and classify the subsidy accordingly but it is recognized that this may be unsatisfactory and it may be felt that longer-term effects should be accounted for. However, as was explained in chapter 6.3 above, this is often difficult and for the costs and earnings analysis, we generally focus our attention on one year as we will see below.

---

[12] This aspect is, however, not discussed further in the Guide but could be an important input into a more in-depth analysis. It should also be noted that the notion "longer-term" in this context is somewhat different from when talking about the longer-term impact of categories 3 and 4 subsidies. Here the impact is noticed in the profit and loss account from the beginning while the longer-term effects of categories 3 and 4 subsidies are generally more implicit.

7.2 Information requirements

With the subsidies classified and assessed so that we have estimates of their values to the industry with regard to revenues and different types of costs, we can proceed to look at these estimates in the context of a costs and earnings analysis. This will require that we have access to income statements of the relevant parts of the industry. In some cases, we may be able to use officially published statements or use financial statistics made available through other surveys or research. However, in most cases we need to consult with the private sector directly. This direct contact is generally recommended as it will also give us the opportunity to verify our findings and ensure that we have understood the impact of subsidies on the particular industry correctly. It should be remembered, though, that it is likely to be a time consuming task, both for our study team and for the respondents. We may also encounter one or several of difficulties common to data collection in this field. Particularly in the informal sector, written accounts are not always kept and income statement as such do thus not exist but have to be estimated through discussions. Moreover, financial information on commercial activities is often considered confidential and operators may be hesitant to cooperate for this reason. Businesses might also fear that the purpose of the study is to set the stage for removing subsidies and hence be unwilling to cooperate. Our fisheries subsidies study may already be limited to a certain segment of the fisheries sector, e.g. the harvesting subsector or one type of aquaculture, but for this part of the study we should probably consider to limit our focus even further given the data requirements. If our study as a whole covers the entire sector, we are likely to want to choose one or a limited number of subsectors for the costs and earnings analysis.

Already in chapter 4.4, it was explained that we need to study fisheries subsidies also from the angle of the industry. It was suggested that we make an inventory of all the operators of the fisheries sector and their activities. This type of information is needed also for the costs and earnings analysis. Firstly, such an inventory would help us to select and sample the segment of the sector that we would like to include in the analysis. Secondly, we need to estimate to what extent different operators are affected by the various subsidies. Depending on the type and scope of their activities, different operators' revenues and costs may be influenced in different ways and at different degrees by the same subsidy. In our assessment of the subsidies, we probably calculated aggregated costs and values but in our more detailed analysis, we need to know how much of this aggregated value that should be apportioned to each segment or to each operator. For example, in Box 5, we saw that, in our invented country Seidisbus, the government operated an investment grants scheme for improving fresh fish storage and transport in the aquaculture subsector at a total value of US$ 2 300 000 in 2000. But there are many different types of aquaculturists in the country and some may have benefited more than others from the subsidy. If the main part of the grants were applied for by the export-oriented brackishwater shrimp producers and our costs and earnings analysis covers only the small-scale rural aquaculture, only a small portion of the total value of the grants scheme is likely to be of relevance to us.

Depending on the data available, this division of the total value of the subsidy between different beneficiaries may be done in different ways. In some cases, we may have detailed information on, for example, exactly who has received a certain grant. In other cases, we need to construct a distribution index to allocate the aggregated value in a more approximate way. We may need to use a sample population and make assumptions about how the results can be extrapolated. If we have a clear view of the economic structure of the sector, this work is greatly facilitated.

Box 18: Structure of the fisheries sector - An example

An inventory and description of the fisheries industry in our country Seidisbus gave the following summary information:

Type of commercial activity	Total sales volume	Unit	Sales value (M US$)	Number of operators / firms	Number of employees	Remarks
INPUT AND SUPPORT INDUSTRY:						
Boat builders (private)	30	Arti-sanal craft	1.5	3	20	
Ship wharf (state-owned)	2	Fishing vessels	20	1	200	
Fishing gear importers (nets, engines, etc)	n/a	n/a	10	25	55	
Repair workshops / artisanal	n/a	n/a	5	20	30	
Repair workshops/ major	n/a	n/a	10	2	15	
Fish feed producers	5 000 000	tonnes	10	2	20	
Various support services (informal sector)					200	Estimate
TOTAL SUBSECTOR			*56.5*	*53*	*540*	
CAPTURE FISHERIES (MARINE):						
Artisanal fishers	20 000	tonnes	30	500	3 500	800 boats
Semi-industrial fleet	10 000	tonnes	10	20	250	30 boats, mainly small pelagics
Shrimp trawlers	5 000	tonnes	35	6	120	8 boats
TOTAL SUBSECTOR	*35 000*	*tonnes*	*75*	*526*	*3 870*	
AQUACULTURE:						
Small-scale farmers (fresh water, < 3 ha / pond)	20 000	tonnes	20	5 000	10 000	Often household based; carps
Prawn farmers	1 000	tonnes	8	12	30	
TOTAL SUBSECTOR	*21 000*	*tonnes*	*28*	*5 012*	*10 030*	
PROCESSING:						
Small artisanal plants	25 000	tonnes	75	200	400	
Industrial plants	30 000	tonnes	300	10	500	
TOTAL SUBSECTOR	*55 000*	*tonnes*	*375*	*210*	*900*	
MARKETING AND DISTRIBUTION:						
Exporters	10 000	tonnes	150	3	20	
Retailers in local markets	15 000	tonnes	75	1 000	1 200	Estimates
TOTAL SUBSECTOR	*25 000*	*tonnes*	*225*	*1 003*	*1 220*	
GRAND TOTAL				*6 804*	*16 560*	

NB. All figures are fictive and may not correspond to real market prices or values.

7.3   Income statement with and without fisheries subsidies

Once we have the information – by operator or sector segment depending on how we have selected the focus of our analysis – we should organize the subsidy values and income statement data into a format allowing us to calculate a profit and loss account with subsides, i.e. representing the actual or current situation, and one in which we have removed the subsidies.

Regarding revenues and the variable, labour and fixed costs – if following the structure proposed above for the subsidy classification – there is not likely to be much difficulty once we have obtained the relevant figures from the industry. However, with regard to depreciation and interest costs, we may want to standardize our values.

In the fishing industry, depreciation costs are commonly an important item in the financial accounts owing to the importance of vessel investments. Tax regulations for how to account for these depreciation costs vary considerably between different countries. Interest is another highly variable cost in a profit and loss account as it depends on the level of loans and does not include the capital opportunity cost. If we would like to compare the results of the analysis, and the effect of subsidies, between different operators or groups of operators, a standardized method for valuing depreciation and interest costs would be needed. This would also assist us in understanding the economic performance of the industry better. It is hence suggested that the actual or current income statements are recalculated, using standardized methods for valuating depreciation and interest costs[13].

With regard to calculating the depreciation cost for a fishing vessel, the basis is the factual replacement value of the vessel. This could equal the current building costs of a similar new vessel or be the price of a second-hand vessel, if this is the most likely scenario for replacing vessels according to local commercial practices. The annual depreciation value is then calculated as the replacement value divided by the expected economic life span. The expected economic life span should also be calculated based on knowledge of local conditions.

With regard to the interest cost, the nominal book value of the vessel – calculated in accordance with the above-described methodology – should be used as the value on which to impute interest cost. The interest rate used can be the rate for government bonds adjusted for inflation or any other measurement considered being an appropriate measurement of real interest rates.

So far we have only mentioned depreciation and interest costs related to vessels. However, the Guide suggests that the same principles are applied to other subsectors and that profit and loss accounts including adjusted depreciation and interest costs are calculated – to the extent possible – also for companies with other activities involving important capital investments, e.g. processing plants.

An example of a costs and earnings analysis is presented in Box 19. It is based on the information given on our invented country Seidisbus in the various examples from chapter 6.2 and the sample sector inventory in Box 18.

---

[13] The methodology for uniform depreciation and interest cost calculations for fishing vessels was developed by the Department of Fisheries of the Agriculture Economics Research Institute in the Hague (Davidse *et al* 1993) and has also been used in cost and earnings analyses of fishing units carried out by FAO (Lery, Prado and Tietze1999 and Tietze *et al* 2001).

Box 19: Costs and earnings analysis - An example

Within the framework of the fisheries subsidies study in Seidisbus, a costs and earnings analysis was carried out for the shrimp fishery. In 2000, there were six companies operating a total of eight boats in this fishery. Income statements were obtained from four of the companies, covering six of the boats. It was assumed that the remaining two companies operated along similar lines to those interviewed and the data obtained was extrapolated to establish a profit and loss account for the shrimp fishery as a whole. All eight vessels in the fishery were of the same type and size although of very different age; the newest had just been taken into service while the oldest had been operating for nearly 25 years. The average age of the fleet was estimated at 8 years. The cost of a new vessel was estimated to be about US$ 10 000 000 based on information from the national ship wharf. However, there is also an important second-hand market in the region and the average value of the vessels in the fleet was estimated to be 6 000 000 with a life span of 15 years. Accordingly, the total current value of the fleet is US$ 48 000 000. The commercial interest usually charged for this type of investment was 15% and the loan period would generally be the same as the economic life span of the investment, i.e. 15 years in this case, with payments due at the end of each year.

With regard to subsidies, many of the examples quoted in section 6 – and summarized in

Figure 9 – were relevant to the shrimp fishery in Seidisbus. The exceptions were the investment grant program (Box 5) which was only relevant to the aquaculture subsector, the provision of landing sites along the coast for the artisanal fishers (Box 13), the membership in the regional fisheries committee dealing with the management of small pelagic species (Box 15) and the extra costs for new TEDs which were already being used by the fleet (Box 16).

The **income guarantee scheme** (**Category 1 subsidy**) benefited the fishers working onboard the shrimp trawlers. However, there was no information on the amounts having been paid out to individual fishers and it was hence assumed that the scheme had benefited fishers in the semi-industrial and the shrimp fishery fleets equally:
> 120 (employees shrimp fleet) divided by 370 (total employees semi-industrial and shrimp fleets) multiplied by 450 000 (industry value of subsidy) = 145 900.

All six companies targeted the export market and had their own marketing and distribution structure. Shrimp exports represented 90% of the total value of fish exports in 2000. Four of the companies had participated in the 2000/2001 trade fair organized by the **Export Council** (**Category 2 subsidy**).
> 90% of 75 000 (fisheries' share of Export Council budget) plus 4 / 30 (share of shrimp industry participants in trade fair) of 12 000 (fisheries' share or trade fair costs) = 69 100.

For the **fuel tax rebate** (**Category 2**), there were records of the recipients of the reimbursements. The shrimp fishery fleets had received a total of 550 000 under the scheme.

FAO, and in particular the marine fisheries management project (**Category 2**), was important to the shrimp fishery, probably more so than to many other parts of the sector. It was believed that it would be fair to assign 75% of the industry value to the shrimp fleet: 75% of 201 750 = 151 300.

The **free access subsidy** (**Category 4**) affected the shrimp fishery proportionally to the value of their landings, i.e. 4% of 35 million = 1 400 000.

AGGREGATED PROFIT AND LOSS ACCOUNT – SHRIMP FLEET (US$) 2000				
Item	Actual: adjusted depreciation and interest costs	Name of subsidies	Amount of subsidy	Account less subsidies
REVENUES				
SALES REVENUES	35 000 000	FAO Export Council	151 300 69 100	34 779 600
OPERATING COSTS				
RUNNING (VARIABLE) COSTS	17 000 000	Fuel rebate	550 000	17 550 000
LABOUR COSTS	5 000 000	Income guarantee	145 900	5 145 900
FIXED COSTS	3 000 000	Free access	1 400 000	4 400 000
GROSS CASH FLOW	*10 000 000*			*7 683 700*
CAPITAL AND FINANCIAL EXPENSES				
DEPRECIATION	3 200 000			3 200 000
INTEREST COSTS	500 000			500 000
PROFIT OR LOSS BEFORE TAX / TOTAL SUBSIDIES	*6 300 000*		*2 316 300*	*3 983 700*
TAX				
CORPORATE INCOME TAX (15%)	945 000			597 555
PROFIT OR LOSS AFTER TAX	*5 355 000*			*3 386 125*

8   COMPARATIVE ANALYSIS

8.1   Relative importance of costs and values

The estimates of fisheries subsidies that we made above represent important information but to better assess their significance the values need to be compared with something. The industry value of a subsidy could, for example, be put in relation to the total sales value for the part of the industry it affects, or the total government expenditure on fisheries subsidies could be expressed as a percentage of the total added value created by the fisheries sector[14] and compared with similar ratios for other sectors.

Which ratios that should be calculated depend of course on the objective of the analysis, e.g., should the fisheries subsidies be compared with other sectors of the economy or with fisheries in other countries, or should the development – increases or decreases in different categories of fisheries subsidies – over time be measured? Some examples of ratios that could be calculated are listed below. The ratios can either be calculated for the fisheries sector as a whole or for different subsectors or groups of firms, depending on the scope and objective of our study.

- *Government expenses (revenues)*
  - ➢ Government costs (revenues) divided by the number of employees in the fisheries sector, selected subsectors or groups of operators.
  - ➢ Government costs (revenues) divided by the value added created by the sector or subsector.
  - ➢ Government costs (revenues) divided by the value of production (turnover) of fisheries industry or part of it.
  - ➢ Government costs (revenues) divided by ex-vessel value of landed fish.

- *Change in industry profits*
  - ➢ The industry value of the subsidies divided by the total profit/loss (before or after tax) of the fisheries industry, selected subsectors or groups of operators.
  - ➢ The industry value divided by the ex-vessel value of landed fish.
  - ➢ The industry value (change in profits) divided by the value added created by the fisheries sector, selected subsectors or groups of operators.

Box 20: Ratios - An example

In Seidisbus, the following ratios are calculated in the fisheries subsidies study:

1. Government cost (all subsidies) divided by the total number of employees in the fisheries sector:
5 353 750 (from Figure 9) / 16 560 (from Box 18) = US$ 323 per employee.
2. Government cost (only Categories 1 and 2 subsidies) divided by the total number of employees in the fisheries sector: 2 418 750 (from Figure 9: 5 353 750 – 2 900 000 – 35 000) / 16 560 = US$ 146 per employee.
3. Government cost (excluding subsidies for aquaculture) divided by the ex-vessel value of catches:
4 583 750 (from Figure 9: 5 353 750 – 770 000) / 75 000 000 (from Box 18) = 6%.
4. Industry value (all subsidies) divided by the ex-vessel value of catches:
7 053 750 (from Figure 9) / 75 000 000 = 9%.
5. Industry value (subsidies only for shrimp fishery) divided by profits before tax of the shrimp fleet:
2 316 300 (from Box 19) / 6 300 000 (from Box 19) = 37%.

8.2 Financial ratios

In addition to the more general ratios discussed above, we may also want to make further use of the results of the costs and earnings analysis. Based on the calculations made on the income statements – discussed in chapter 7 above – we can calculate financial ratios and in this way evaluate the economic performance with and without subsidies. Depending on our sample size and the number of subsectors that we have included in the costs and earnings analysis, average ratios for different parts of the industry can be estimated and assessed. Some of the ratios to calculate could include:

- Gross margin
- Profit margin (return on sales)
- Internal Rate of Return (IRR)
- Return on investments

It would also be interesting to examine the change in financial strength and solvency ratios but as the longer-term impact of the subsidies on the firm is not known, this would be difficult to do in any reliable way. The financial strength and solvency ratios are based on information from the balance sheet and in order to make any meaningful assessment, the balance sheet would need to be adjusted for subsidies in the same way as the profit and loss account. The latter is a shorter-term reflection of the business and it is easier to make adjustments with an acceptable level of reliability. The balance sheet is the long-term account of the business's transactions. To adjust the balance sheet for the effects of subsidies would involve, in addition to analysing the history of the direct effects of subsidies, speculations with regard to overall investment and business decisions triggered by the indirect effects of subsidies in the past. In fact, in line with this discussion, also the last profitability ratio suggested above, i.e., *return on investments*, could be questioned with regard to its reliability as it uses total assets – a balance sheet item – as the denominator.

Box 21: Financial ratios - An example

Using the information for the shrimp fishery in Seidisbus (Box 19), the following financial ratios can be estimated:

1. **Return on sales**
   Actual account: 6 800 000 (net income before interest expenses: 6 300 000 + 500 000) divided by 35 000 000 (sales) = <u>19%</u>.
   Account less subsidies: 4 483 700 (net income before interest expenses: 3 983 700 + 500 000) / 34 779 600 (sales) = <u>13%</u>.
2. **Return on investment**
   Actual account: 6 800 000 (net income before interest expenses: 6 300 000 + 500 000) divided by 48 000 000 (book value of total assets assumed to equal current replacement value of vessels) = <u>14%</u>.
   Account less subsidies: 4 483 700 (net income before interest expenses: 3 983 700 + 500 000) / 48 000 000 (book value of total assets assumed to equal current replacement value of vessels) = <u>9%</u>.

---

[14] The added value created by the fisheries sector is often referred to as the GDP of the fisheries sector.

9   WHAT TO INCLUDE IN A SUBSIDY DESCRIPTION

So far in this Guide, we have been concerned mainly with the classification of fisheries subsidies and assessments of their costs and values. These analyses of course have to be based on knowledge about the subsidies, including both quantitative and qualitative information. In a report on fisheries subsidies, it would be advisable to organize this information in a standardized format. By recording detailed descriptions of the various subsidies, the information is readily available and can be used as an input into further analyses or for reporting to, for example, the WTO[15]. Accordingly, a format for how to describe fisheries subsidies is proposed in Table 2 below. It is suggested that each subsidy is described separately according to this checklist.

Table 2: Description of fisheries subsidies – a checklist

NAME OF THE SUBSIDY
FORM OF SUBSIDY AND SHORT DESCRIPTION OF ITS FUNCTION
CLASSIFICATION: CATEGORY 1, 2, 3 or 4 *(see chapter 5 of the Guide)*
HOW IS THE PROFITABILITY OF THE INDUSTRY AFFECTED BY THE SUBSIDY *(revenue-enhancing / cost-reducing etc: see chapter 7 of the Guide)*
PERIOD OF IMPLEMENTATION *(date of introduction [month/year] to end-date [month/year] – or on-going)*
POLICY OBJECTIVE AND PURPOSE OF THE SUBSIDY *(short description of why the subsidy has been introduced and what its economic / social / development / environmental objective is)*
RESPONSIBLE MINISTRY / DEPARTMENT / AUTHORITY / ORGANIZATION AND LEGISLATION UNDER WHICH THE SUBSIDY IS GRANTED, IF APPLICABLE
FUNDING OF THE SUBSIDY *(fully government funded or with contributions from the industry)*
COVERAGE AND TARGET RECIPIENTS (to whom is the subsidy provided) • APPLIES TO WHICH (SUB)SECTORS • APPLIES TO WHICH GEOGRAPHICAL REGIONS • WHO ARE THE RECIPIENTS • WHAT ARE THE CRITERIA FOR RECEIVING THE SUBSIDY
DESCRIPTION OF THE MECHANISM BY WHICH THE SUBSIDY IMPLEMENTED (process through which the benefits/disadvantages are transferred or created, process through which the recipients learn about/apply for/receive the subsidy, etc.)
REVIEW OF ANY SIDE EFFECTS OR EXTERNALITIES CAUSED BY THE SUBSIDY *(specify indirect beneficiaries or who it affects and how)*
ASSESSMENT OF THE SUBSIDY • GOVERNMENT COST (REVENUE) OF THE SUBSIDY • INDUSTRY VALUE OF THE SUBSIDY *(specify the year or period of the assessment and give details of the calculations)*
ANY OTHER RELEVANT INFORMATION *(e.g. ratios, statistical data for assessing the subsidy in a particular context such, for example, trade, etc.)*

---

[15] The format for subsidies description suggested by the Guide is partly based on the information to be provided in WTO's questionnaire for subsidy notifications. However, the statistical data required in the questionnaire needed for the assessment of the trade effects of the subsidy are not explicitly covered here.

10 REPORTING ON SUBSIDIES

10.1 The study report

In this last section of the Guide, we will briefly review some aspects with regard to reporting on fisheries subsidies. Examples and suggestion for how to organize the information collected and our results have already been included in different parts of the Guide, e.g. Figure 6 and Figure 9 on listing fisheries subsidies, chapter 7 on making an inventory of the fisheries industry and on costs and earnings analyses, and chapter 9 on descriptions of fisheries subsidies.

If our fisheries subsidies study is being carried out on the request from, for example, a government authority with a particular objective and terms of reference, or it is part of larger research task, we may already have an outline for how our report should look like. If this is not the case, the suggested outline in Table 3 may give some useful ideas.

Table 3: Tentative outline for the final report of a fisheries subsidies study

No	Chapter	Chapter in Guide
1 1.1	INTRODUCTION Background and purpose • *Application of the study in the national context*	1-2
1.2	Methodologies • *Basic concepts and main principles* • *Survey and data collection methodologies*	3-10
1.3	Limitations • *Precisions of the coverage of the study with regard to, for example, time frame, subsectors, geographical areas or subsidy categories* • *Description and explanations with regard to particular problems encountered, e.g., methodologies or data availability* • *Appraisal of the validity and reliability of the study results*	2
2	THE MACROECONOMIC FRAMEWORK • *Brief description of the main economic and policy aspects*	4
3	THE FISHERIES SECTOR • *Brief description of the fisheries sector* • *Inventory of the different economic activities of the different subsectors or groups of operators*	4 & 7
4 4.1	PRESENTATION OF RESULTS Description and assessment of subsidies • *Categories 1-4* • *Government cost (revenue) and industry value*	4-6 & 9
4.2	Costs and earnings analysis	7
4.3	Comparative analysis and ratios	8
5	CONCLUSIONS AND RECOMMENDATIONS	10
Bibliography		
Appendices	• Terms of reference for the study • Details of the results of the institutional survey and the fisheries sector review • Details on methodologies and assumptions • Detailed descriptions of the investigated subsidies	

It is recommended that we put most our efforts into chapter 4 "Presentation of results". It should be remembered that chapters 2 and 3 are not meant to be the core of the report and even though it is usually relatively easy to include a large amount of interesting information here, they are better kept to a minimum of issues, necessary for giving the context and framework of the subsidy identification and assessments. If we want to include more details, on, for example, the results of the institutional survey or the fisheries sector review, this may be better done in an appendix.

The exact sub-chapters to include in chapter 4 "Presentation of results" will of course depend on the scope of our study but, most likely, a substantial part of the report will be on the "Description and assessment of subsidies". Here we may want to present our findings in a summary listing (see Figure 6 and Figure 9) in addition to giving descriptions of the subsidies as well as explanations regarding the assumptions made for their assessment in the text. We may opt for including additional information on the subsidies (according to Table 2) and more details on the methodologies in an appendix to the report.

10.2 Summary of the Guide

Fisheries subsidies studies can have different objectives and scopes and thus consist of different components. In this Guide, we have discussed those that are felt to be the most important ones. It is of course up to the user of the Guide to decide which parts of the Guide that are most relevant to his or her particular study. Figure 10 summarizes the different components as they have been presented in the Guide and stresses the importance of collecting information both from the public sector and the fisheries industry.

Figure 10: Summary of the Guide's components

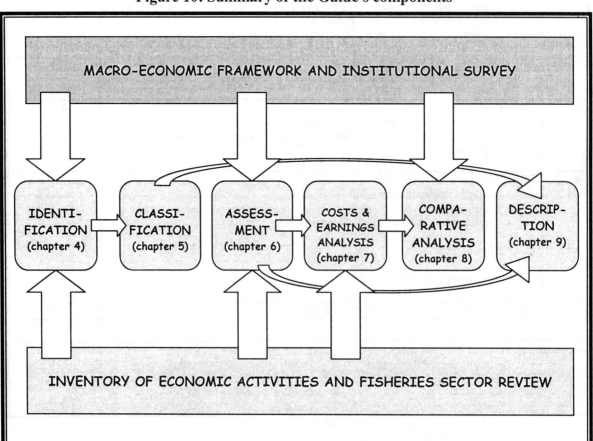

BIBLIOGRAPHY

Davidse, W.P., Cormack, K., Oakeshott, E, Frost, H. and C. Jensen. 1993. Costs and earnings of fishing fleets in four EC countries: Calculated on an uniform basis for the development of sectoral fleet models. Ondersoekverslag, Landbouw Economisch Instituut, The Hague, 202 p.

EEC, 1997. Council Regulation (EC) No 189/97 of 26 September 1997 imposing a definitive countervailing duty on farmed Atlantic salmon originating in Norway. Official Journal L267, 30/09/1997. P. 19-44.

Encyclopaedia Britannica. 2001. Deluxe Edition CD-ROME.

FAO. Fisheries Glossary (available at www.fao.org).

FAO. 2000a. Report on the Expert consultation on economic incentives and responsible fisheries. Rome, Italy, 28 November - 1 December 2000. FAO Fisheries Report. No. 638. Rome, FAO. 25 p.

FAO. 2000b. The state of world fisheries and aquaculture 2000. FAO Fisheries Department, Rome.

Flaaten, O. and Wallis, P., 2000. Government financial transfers to fishing industries in OECD countries. Expert consultation on economic incentives and responsible fisheries. Rome, Italy, 28 November - 1 December 2000. FI:EIRF/2000/Documents made available 6. 13 p.

Hannesson, R. 2000. Management costs in fisheries and their recovery: some principles. Expert consultation on economic incentives and responsible fisheries. Rome, Italy, 28 November - 1 December 2000. FI:EIRF/2000/Documents made available 5 (1).

Lery, J.-M., Prado, J., and U. Tietze. 1999. Economic viability of marine capture fisheries - Findings of a global study and an interregional workshop. FAO Fisheris Technical Paper No. 377. FAO, Rome, 130 p.

Milazzo, M. 1998. Subsidies in world fisheries - a re-examination. World Bank Technical Paper No. 406. The International Bank for Reconstruction and Development/World Bank, Washington. 86 p.

Munk, K.J., and M. Motzfeldt. 1993. Definition and measurement of trade distortion for the fishing industry. Commission of the European Communities, Brussels, 56 p.

OECD. 1993. Economic assistance to the fishing industry: Observations and findings. OECD, Paris, 14 p.

OECD. 2000. Transition to responsible fisheries - Economic and policy implications. OECD, Paris, 272 p.

PricewaterhouseCoopers LLP. 2000. Study into the nature and extent of subsidies in the fisheries sector of APEC Member Economies CTI 07/99T Draft Report - End module synthesis. Prepared for Fisheries Working Group, Asia Pacific Economic Co-operation

(APEC). Expert consultation on economic incentives and responsible fisheries. Rome, Italy, 28 November - 1 December 2000. FI:EIRF/2000/Documents made available 7.

Schrank, W.E., and W.R. Keithly Jr. 1999. Thalassorama - The concept of subsidies. Marine Resource Economics, XIV, p. 151-164.

Silvis, H.J., and C.P.C.M van der Hamsvoort. 1996. The AMS in agricultural trade negotiations: A review. Food Policy, Vol. 21, No. 6, pp. 527-539.

Tietze, U; Prado, J; Le Ry, J-M; Lasch, R. 2001. Techno-Economic Performance of Marine Capture Fisheries and the Role of Economic Incentives, Value Addition and Changes of Fleet Structure. Findings of a global study and an inter-regional workshop
FAO Fisheries Technical Paper No 421. Rome, FAO.

Wallis, P., and O. Flaaten. 2000. Fisheries management costs: Concepts and studies. Paper presented at IIFET 2000. (www.oecd.org/agr/fish/index.htm)

WTO Committee of Trade and Environment. 1999/2000. Reports of meetings and national submissions (February 1999 to October 2000). Expert consultation on economic incentives and responsible fisheries. Rome, Italy, 28 November - 1 December 2000. FI:EIRF/2000/Documents made available 4.

INDEX

APPENDIX I:

GLOSSARY

Ad valorem tax
A tax that is a percentage of the selling price. An example of an ad valorem tax would be the Value Added Tax (VAT).

Cash flow
A record of an organization's liquidity, i.e. cash income and cash payments in a given period of time

Depreciation
The decline in value of an asset due to wear, age or technological obsolescence over its economic life span. Depreciation applies both to tangible assets, such as inventory or machinery, as well as intangible assets, e.g. copyrights, licences or leases. For accounting and tax purposes, standardized methodologies for calculating annual depreciation costs are often used that do not necessarily reflect the true economic depreciation.

Direct Foreign Investment
Acquisition or construction of physical capital by a firm from one country in another country.

Equity capital
The equity of company is the funds that have been invested in the company by its owners. Government equity capital occurs when the government is the investor.

Exchange rate
Rate at which one currency may be converted into another. Also called rate of exchange or foreign exchange rate or currency exchange rate.

Fisheries industry (as opposed to public administration)
All productive sub-sectors of the fisheries and aquaculture sector comprising recreational, subsistence and commercial fishing, and including the harvesting, processing, and marketing sectors.

Fixed asset
A long-term asset that is not expected to be converted into cash in the current or upcoming fiscal year. Fixed assets can be both tangible assets, such as fishing vessels, processing plants and real estate, as well as intangible assets, e.g. goodwill, patents and share holdings.

Fixed costs
Production costs that do not vary with the output quantity. Fixed costs could include building or office rent and marketing costs.

Gross margin
Gross margin is a financial ratio describing the gross profit. It is expressed as a percentage and is calculated as the gross income (operating income before depreciation) divided by total sales.

Inter-bank offer rate
The interest rate that the largest international banks charge each other for loans.

Internal rate of return (IRR)
IRR is a financial ratio expressing the net present value of expected future cash flows as a percentage of an investment. In financial analysis, IRR can be used for evaluating the return on an investment and to compare different investment options.

ITQ (Individual Transferable Quota)
A type of quota (a part of a Total Allowable Catch) allocated to individual fishermen or vessel owners and which can be sold to others.

Market prices
In economic terms, the market price is the price at which the market is in equilibrium, i.e. at which supply and demand converge. In more general terms, the market price is the price at which products and services are generally available to consumers in a market economy.

Opportunity costs
The benefit foregone by using a scarce resource for one purpose instead for its next best alternative.

Overhead (costs)
The ongoing administrative expenses of a business, such as rent, utilities, and insurance.

Profit margin
Profit margin is a profit ratio expressing the profit as a percentage of total sales. It is calculated by dividing income before extraordinary items and interest expenses by total sales.

Shadow prices
Any distortion of a free market price that is made in order to reflect the real scarcity value of goods or services, including labour. If no market price exists, this is the unobserved hidden or implicit price that is derived through inferences.

Usury rate
An illegally high interest rate on a loan.

Value-added
The value that has been added to a good through production or processing, i.e., the value of the final good minus the costs for buying raw materials and intermediate goods.

Variable costs
Production costs that vary with the quantity of output. If output increases, then the variable costs will increase.

APPENDIX II:

OTHER SUBSIDY CLASSIFICATIONS

There are many ways of classifying subsidies and also many possible subcategories available. Some of the main aspects found in the literature according to which subsidies can be classified are:

- *Modalities*
  Classification according to how the subsidy works, i.e., what mechanism it has in the fisheries sector. In their report on subsidies and support programmes in the APEC countries, PricewaterhouseCoopers (2000, page 8) has developed a list of six modality categories, i.e.:
  - ➢ Direct assistance to fishers and fish workers
  - ➢ Lending support programmes
  - ➢ Tax preferences and insurance support programmes
  - ➢ Capital and infrastructure support programmes
  - ➢ Marketing and price support programmes
  - ➢ Fisheries management and conservation programmes

  OECD also classifies subsidies (GFTs) according to how the transfers are implemented, i.e., as Market price support, Direct payments, Cost reducing transfers or as General services. The latter covers the subcategories fisheries management, enforcement and research (OECD 2000).

- *Application*
  Classification according to where in the fisheries sector the subsidy exists. PricewaterhouseCoopers (2000, page 9) defines three subsectors, i.e., Capture fisheries, Aquaculture and Fish processing. In cases where the industry is vertically integrated to a high degree, it may at times be difficult to clearly define the limits between the different subsectors.

- *Origin and specificity*
  Classification according to which government body is funding the subsidy – a fishery specific department or institution such as the Ministry of Fisheries, or one not directly related to fisheries – and whether the subsidy is specific for the fisheries sector or available also to other sectors. Subsidies can also be divided into local, national or regional subsidies. Milazzo (1998) reports on two types of "cross-sectoral subsidies": aid to shipbuilding and infrastructure development. Support to an underdeveloped geographic region, such as the Norwegian Industrial and Regional Development Fund, is an example of a subsidy benefiting the fisheries sector even though not targeting it directly (EEC 1997). A change in monetary policies, e.g., of interest rates, or in tax rates also affects the fisheries industry even if the intervention is general and originates outside the fisheries sector (Schrank and Keithly Jr. 1999).

- *Small scale vs. Large scale*
  Classification according to the monetary importance of the subsidy, either with regard to the total public expenditure or the benefits to single operators (PricewaterhouseCoopers 2000).

- *Short- vs. Long-term*
  Classification according to <u>within what time frame the subsidy is affecting the profitability</u> of the industry. Subsidies implying changes in capital usually mean long-term effects. However, the issue is complex and, for example, a scheme subsidising investment in fishing vessels will have a long-term effect on the profitability of the industry since it implies a change in capital. At the same time, it is known that with an increasing total fishing capacity, the rents from the fishery – and hence its profitability – will eventually diminish and in a further perspective the impact of the subsidy on profitability may be negative (Schrank and Keithly Jr. 1999). Moreover, subsidies are likely to have more implicit effects on efficiency in general and short-term effects on profits will over time translate into the overall economic sustainability of the activity.

- *Budgetary vs. Non-budgetary*
  Classification according to <u>whether the subsidy is identifiable in the Government budget</u>, e.g., the budget of a fisheries agency or department, or un-/under-budgeted, for example subsidized lending or tax preferences. This latter category may also include subsidies from non-fisheries agencies (Milazzo 1998).

- *"Normal" subsidies vs. Conservation subsidies*
  Classification according to <u>whether the subsidies tend to increase production, e.g., the harvesting capacity, or whether they favourably affect the environment,</u> aiming at decreasing fishing operations and enhancing the resource base. The former are often called "bad" subsidies while the latter are commonly considered to be "good" (Milazzo 1998).

- *Positive vs. Negative subsidies*
  Classification according to whether it is a positive <u>subsidy that tend to increase the industry's profitability, e.g., a grant or a loan guarantee, or a negative subsidy reducing profits, e.g., taxes</u>. It should be noted, though, that a subsidy that is negative to the fishing industry would be expected to be positive to society as a whole through positive effects accruing to other sectors. Likewise, externalities resulting from subsidies in other sectors can be negative subsidies for the fisheries industry (Schrank and Keithly Jr. 1999). Individual negative and positive subsidies sometimes cancel each other out. For example, a government levy on landed fish could be classified as a negative subsidy but if it finances a fish price support scheme of which the benefits accrue to the fishers paying the levy, the two programmes together constitute a self-financing activity rather than subsidies. Still, the government regulations supporting the activity can be classified as a subsidy since this is a government intervention affecting the profitability of the industry.

- *Cost reducing vs. Income increasing*
  Classification according to <u>how the subsidy influences the profitability of the industry</u>. In a communication to the WTO, the United States differentiated between Subsidies that reduce capital (fixed) and operating (variable) costs, and Subsidies that support incomes and prices (WTO Committee of Trade and Environment, 1999/2000). This classification can be further broken down and subsidies classified according to what type of earnings and costs that are affected by the subsidy.

APPENDIX III:

MORE EXAMPLES OF POSSIBLE SUBSIDIES OF DIFFERENT CATEGORIES[16]

Category 1

Grants to purchase new or old vessels, or to modernize
Income support, unemployment insurance and income guarantee payments
Vessel decommissioning payments
Licence, permit and quota buyouts and retirement grants
Compensation for closed or reduced seasons
Gear conflict compensation programmes
Disaster relief payments to fishers
Equity infusions to fish processing, harvesting or aquaculture firms by governments
Price support payments to fishers
Grants to small fisheries and direct aid to participants in specific fisheries
Grants to establish joint ventures

Support to improve economic efficiency
Grants for safety equipment
Direct export incentives
Grants for retraining fishers for other industries
Bad weather unemployment compensation schemes
Taxes *(negative)*
Import/export duties *(negative)*
Compensation for damages
Investment grants for pond construction
Grants for temporarily withdrawing fishing vessels
Vacation support payments
Payments to reduce accounting costs
Matching contributions for private sector investment
Transport subsidies

Category 2

Government funded health programmes specific to fisheries
Payments to foreign governments to secure access to fishing grounds
Fishery-specific infrastructure, e.g. fish markets, landing sites and ports
Provision of bait services
Gear development
Support to community based management, regional development and producer organizations
Fuel tax exemptions for vessel fuel
Sales tax exemptions
Special income tax deductions for fishers
Tax exemptions for deep-sea fisheries
Deferred tax programmes
Investment tax credits
Loans made on favourable terms
Government guarantees of bank loans
Fishers' insurance programmes or subsidized insurance
Market promotion programmes
Input and output regulations
Support to consultative groups and mechanisms
Inspection and certification services
Training and extension services
Provision of seeds and feed for aquaculture
Nationality and residence requirements for company officials/managers and crew

Government funded research and development programmes
Reduced charges by government agencies
Sales of commodities to fishers at less than market price
Information collection, analysis and dissemination
Promotion and development of fisheries
Exploratory fishing and gear development
Fisheries enhancement including support for artificial reefs
Research on deep-sea fishing
International fisheries cooperation
Market interventions
Regional development programmes
Tariffs and tariff quotas
Import quotas
Waivers of import duties
Price support systems
Landing bans
Prohibitions on foreign direct investment
Fisheries management (unrecovered costs)
Promotion of fish consumption
Free trade zones
Market research
Ownership restrictions
Allocation of catch quotas only to national fishers

Category 3

Hatchery and fish habitat programmes
Environmental regulations
Enhancement of the fisheries community environment

Chemical and drugs regulations for aquaculture
Production and feed quota schemes in aquaculture
Licence requirements for fish farming
Veterinary surveillance requirements for aquaculture

[16] From FAO Fisheries Department / FIPP (internal working document) and prototype studies.

Technology transfers Protection of marine areas Gear regulations (e.g. TEDs) Food safety and hygiene regulations	Regulations with regard to the escape of fish in aquaculture Record keeping and reporting requirements
Category 4 Free or below market price resource access Lack of implementation of fish quality standards Fisheries registration fees not collected Non-enforcement of existing regulations Lack of pollution control	No requirement of certificate of competence or fisherman's licence Use of free public services, e.g. water; sewerage services, for fishers